Explorations *in* **GLASS**

Explorations *in* **GLASS**

S. DONALD STOOKEY

An Autobiography

With a Foreword by George H. Beall

Published by
The American Ceramic Society
735 Ceramic Place
Westerville, Ohio 43081 USA

The American Ceramic Society
735 Ceramic Place
Westerville, Ohio 43081

03 02 01 00 4 3 2 1

ISBN: 1-57498-124-2

Cover images, clockwise starting from top left: FOTOFORM® glass honeycomb, FOTOFORM® "spider web," FOTALITE® dinner plate, and FOTOFORM® glass lace, all courtesy of Corning Incorporated Department of Archives and Records Management, Corning, New York.

For more information on ordering titles published by The American Ceramic Society or to request a publications catalog, please call (614) 794-5890 or visit our online bookstore at <www.ceramics.org>.

Contents

Foreword

Dr. S. Donald Stookey is a true icon at Corning Incorporated (formerly Corning Glass Works) where he enjoyed a remarkable 47-year career. Don, as he is affectionately known by his many friends, is now retired and living in Florida, but follows scientific progress at Corning with great interest.

Don Stookey is best known as a great inventive scientist, but is also a fine family man and warm human being, as those of us fortunate enough to work for him in our formative years can enthusiastically attest. Not only did Don challenge young scientists to stretch or redirect their conceptual thinking, but he always took a great interest in their families and recreational pastimes. No job-related or personal problem they encountered was beyond his constructive concern. He was a quiet strength in the small town of Corning, New York, generously supporting community causes, especially elderly care.

Don is a true master of invention, having a special insight into the mystery as well as the technical behavior of glass. Nothing was impossible in Don's perceptive imagination; but, more importantly, he displayed just the right combination of technical skills, gritty persistence, and personal credibility to be able to assemble the resources for successful execution of his ideas.

Don's friendship with both Dr. William H. Armistead, longtime Director of Research and Development at Corning, and former Corning CEO and now U.S. Congressman Amory Houghton, Jr., also played a key role in his success. These men were visionary managers who were dedicated to growing the organization through technology. They readily assumed high risks in the commercial development of Don's glass-ceramic and photochromic glass inventions, as well as in the difficult early years of fiber-optic communications, now Corning's major business.

Although a modest man who never enjoyed giving oral presentations of his work, Don Stookey has received numerous awards for creative invention, the highlight being the National Medal of Technology, presented to him by President Ronald Reagan in 1986. When I once asked him what in his career he was most proud of, however, he pointed to the factories and associated jobs created from his research.

Don has many disciples and admirers at Corning Incorporated, Alfred University, and throughout the glass and ceramics community. In this book, he presents in narrative style, the story of his professional life, with many lessons for the student of inventive science and engineering. He also shares his unique research philosophy, which does not always coincide with current models of innovation favored by the business community.

I believe the reader will find this short book free-flowing, informative, and packed with interesting and surprising events, certainly not always following general public perceptions of the scientific method and industrial scene.

George H. Beall
Research Fellow
Corning Incorporated

I

Introduction

The fifteen years since this book was first published have seen worldwide growth of exciting research and innovation in the fields related to glass-ceramics. In recognition of this, the books department of The American Ceramic Society has kindlyrecommended an updated reprint, with a change in title from *Journey to the Center of the Crystal Ball* to *Explorations in Glass*.

I have been very pleased to be generally recognized as the father of this new field of glass chemistry, and to see its widespread proliferation. The first glass-ceramics—supersonic radomes, CORNINGWARE, VISIONS and others—have proved themselves useful, and new increasingly sophisticated inventions are showing up in many places.

Therefore it seems reasonable to tell the convoluted story of the way in which photosensitive glasses and then glass-ceramics were originated.

In looking back, it becomes apparent that a common denominator in all of these new materials is that—in addition to the special characteristics of each—all owe unique properties to the infinitesimal sizes of their crystalline (or precrystalline) particles. Since current scientific literature shows a good deal of interest in making and studying such "nanosized" particles, I have added an epilogue describing those found in photosensitive glasses and glass-ceramics. They do indeed have interesting properties.

II

My Introduction to Corning Glass Works

In the spring of 1940, I was in the final throes of obtaining a Ph.D. degree in physical chemistry at M.I.T. and looking for a job. Most of the top scholars had already been snapped up by Du Pont and Eastman Kodak, although the country was still in a depression. I was not as disappointed as I should have been that the big corporations had not chosen me because I felt that a position with one of those large, well-known laboratories would not allow me to search independently into the real "unknown." As do most young people, I had formulated a long-range goal — to explore far beyond the known fields of science and to discover or create useful new things that no one had ever seen before.

In my boyhood, reading the exploits of early explorers had excited my imagination. Since the physical world was already known, my ambitions turned toward exploration of Mother Nature's inner secrets. It seemed to me that few secrets were left in the conventional sciences of chemistry and physics, although an education in these basic sciences had been a necessary stepping stone. So where would I find the all important door into unknown lands?

My education was complete, and time was running short when my big opportunity came. Dr. J.T. Littleton and Mr. W.C. Taylor, Corning's directors of research and of glass technology, respectively, were at M.I.T. looking for a man. It appeared that there were three openings: one in glass technology, another in writing the history of Corning's research, and the third doing research on "opal glass."

Glass was an unknown field as far as I was concerned. I had learned nothing about it in school and had never heard of Corning Glass Works. Moreover, "opal glass" sounded exotic and mysterious. So I happily agreed to go to Corning for an interview, a decision I have never regretted. The door to a new world of exciting discovery was opening.

As the train approached the village of Corning, which is situated in the scenic Finger Lakes country of upstate New York, the beauty of the forested hills and lakes, so different from the plains of Iowa, kept my spirits light. I must admit that the first sight of Corning was less inspiring. Fuming smokestacks, grimy snow, houses with dark paint, and hilly streets dominated the scene.

My interview day was spent mostly in the factory office of Walter Oakley, a gruff man who scolded every man who entered the office, then grinned and gave a complimentary description of each one after he had left. Mr. Oakley was chief engineer, in charge of all Corning's glass melting operations; he needed another glass technologist to keep the hundreds of production glasses out of trouble. At the end of the day, Dr. Littleton asked me into his office where he told me that Mr. Oakley was offering me a position. I didn't exactly say no but wanted Dr. Littleton to understand my heart was in exploratory research, and I wanted to learn more about opal glass.

A few months later (July 1, 1940), I reported to work starting at $2,500 a year and reported to Dr. Henry Blau who, as it turned out, had just arrived at Corning himself. He had been chief scientist for MacBethEvans Glass Company in Charleroi, Pennsylvania, which had recently merged with Corning. He started to teach me about opal glass, but left Corning six months after my arrival.

At the same time (midwinter 1940–1941), I put a down payment on a new Chevrolet, drove home to Iowa, and married my beautiful Coe College sweetheart, Ruth Watterson. We honeymooned in Minneapolis, then drove back to Corning by way of Niagara Falls, leaving our last fifteen dollars with a justice of the peace who fined me for speeding. Back in Corning, Ruth and I moved into a second-story apartment a hundred yards from the New York Central railroad track, and I reported to a new boss, Harrison Hood. Mr. Hood, Corning's leading glass composition authority, was a tall, modest scientist who generously taught us fledglings the fundamentals of glass chemistry and suggested interesting projects. He never gave orders.

In 1940 Corning was a moderate-sized producer of many glass products. The company was owned and operated by the Houghton family, whose great-great-grandfather had originated it in 1851. Today, still managed by a new generation of Houghtons, Corning Glass Works is a much larger international corporation, diversified beyond glass and ceramics into electronics, health products, and biochemical products.

Corning's research laboratory is one of the two or three earliest American laboratories, along with Eastman Kodak, and the Houghtons' keen interest and long-range vision had been directly responsible for the free rein that permits the researcher to explore as far as he can. I have more and more realized how rare, perhaps unique, is this management principle in American industry. Unfortunately for the prospect of major inventions, most managers think in terms of limited-scope, quick-payoff research projects. This is true of academic research as well. The reader of this book will see that the conventional type of short-range projects would never have resulted in the kinds of major discoveries described herein. Mr. Hood, then, in the Corning tradition, tutored me in the little known science of glass, showed me the factory operations, started me on a small project that would teach me how crystals form in a certain white ("opal") glass, and encouraged me to go my own way.

Part of my introduction was the opportunity to learn what the other researchers were doing and to see the various kinds of glass manufacture going on. It was surprising to see ancient glassblowing arts in one place, and modern high-speed continuous production of light bulbs, tubing, and television tubes in another.

Dr. Bob Dalton showed me products of two experiments he had performed, both of which were to become the direct forerunners of my first steps into the unknown. One of these was a clear dish imprinted with a three-dimensional white opal pattern. Seeing no evidence of the two glasses having been sealed together, I was mystified. Bob explained that the glass was a special type of opal, sodium fluoride thermally opacifiable glass. Cooled from the melting temperature, it remains transparent but, reheated to red heat for a few seconds, sodium fluoride crystals grow until the glass becomes white. Bob had reheated local areas with a tool something like a branding iron to obtain his pattern. Actually, thermally opacifiable opal glasses have been employed in lighting globes and decorative glassware for many years.

Bob's other samples had some similarities with his opal samples. He showed me some pink dishes with darker red patterns and a sheaf of glass fibers, bright red at one end and colorless at the other. These were copper ruby glass. Copper ruby and gold ruby glasses have been well known art glasses since the Middle Ages. Like the thermally opacifiable opals, such glasses are clear and colorless when first cooled, but develop the ruby color when reheated. The new discovery, made by Dalton, was as follows: Exposure to sunlight or to a carbon arc before heat treatment resulted in a darker red color after subsequent heat treatment. Faced with a busy schedule of developing glasses for industry, Bob had put this "plaything" on the shelf and invited me to play with it.

It didn't take long to realize that my dreams of exploring a mysterious unknown world might be coming true! Library research and discussions with elders in the glass factory revealed that opal glasses and ruby glasses had originated before the dawn of chemistry; the formulas were still secrets, and much remained to learn about them.

I learned that "opal glass" is a term used in the glass industry to designate white opaque or translucent glassware generally employed as tableware or lamp globes. The terms probably originated with the resemblance of translucent glass to the gem stone. The opacity is due to light scattering by large numbers of transparent colorless microscopic particles, usually crystals, uniformly dispersed in the glass. The refractive index of the crystals differs from that of the glass, so rays of light are bent many times and scattered in all directions, and the glass appears white. In manufacture, these glasses are clear and colorless liquids in the hot melting tank and become opaque during cooling or reheating as the crystals, like miniature snowflakes, precipitate inside the formed product. "Opal" crystals include calcium and sodium fluoride, calcium or barium phosphate, and other compounds.

Ruby glasses, developed in Germany in the sixteenth century, owe their beautiful reds to dispersion of metal particles such as gold or copper or cadmium sulfoselenide particles, too small to scatter light or to be distinguished with an optical microscope. The color is due to absorption of light in part of the spectrum. These glasses, like the opals, are clear and colorless in the melting tank. They remain colorless when cooled to room temperature and become colored after reheating to red heat (700°–800°C). Ruby glasses are generally employed in decorative artware. The first gold ruby glasses were invented by a German alchemist and were believed to have supernatural curative powers.

Dr. Dalton told me that sunlight had been known to change the color of old windows or bottles (this phenomenon is known as "solarization"), but that solarized glass loses its color when the glass is reheated. What was different about the Dalton discovery? And what were the differences between the chemistry of the copper rubies and the thermally opacifiable glasses? These questions intrigued me. And I began to wonder whether the chemical rules I'd learned in school had any relevance to this fascinating medium, glass.

It wasn't long before I decided my first big research goal would be to try to invent a photosensitive opal glass, one which could develop a white pattern in clear glass after localized exposure to light and subsequent heat treatment. My reasoning was that, win or lose, I'd learn some interesting glass chemistry, and technical success might produce a new process for decorating glass. Well, I did invent photosensitive opal glasses, but it wasn't my first invention, and I had to learn quite a lot of chemistry before success, as we shall see.

When I started work, the country was still in a depression. Dr. J.T. Littleton, a physicist, was director of research, and the majority of the technical staff were also physicists. W.C. Taylor was head of glass technology, with one or two chemists and some physicists. Technical work mostly involved television glass and chill-tempered glass together with trouble-shooting projects in the factory.

For six months my office and laboratory were in a big dusty attic above the batch mixing room for "B" factory, but then I became one of the first tenants in the new laboratory, Building 50. The only experimental glass melting facility was a muffle oven that replaced one of the clay melting pots for hand-shop glassblowing in "B" factory. "Woody," E.A. Wood, a wry veteran glassblower, mixed one-pound batches, put nine batches a day into clay crucibles, and melted the glass at the pot temperature, generally about 1350°C. Then he pressed flats or drew canes and annealed the flats in an oven. We did have an excellent physical measurements laboratory and an analytical chemistry laboratory under Dr. Jim Farncomb.

Photosensitive Glasses

My first research goal was to invent a clear glass capable of developing an opaque white pattern where it had been exposed to light and reheated. Decorative tableware, I thought, would be a good product. Starting with Dalton's thermally opacifiable opal glass composition, I decreased the sodium fluoride in small steps, in a series of crucible melts, hoping to find a glass that would remain clear on reheating except where it had been exposed to ultraviolet light. But I could find no effect of the light at all.

Photosensitive Copper Ruby

I then studied the copper ruby glasses more carefully. It became apparent that, unlike the opal glasses, these glasses contained extra ingredients, such as compounds of tin and antimony. Finally, the explanation came to me. Copper must dissolve as an oxide in the molten glass but, when reacting with the special compounds, it gradually is reduced to the insoluble metallic state as the glass is reheated, developing tiny coloring particles of metal. The photosensitive copper ruby glass, then, contains less of the reducing agent, and requires irradiation to complete the reduction to metal. Such photochemical reactions occur in liquids, but had never been recognized in glass before. Soon I produced real photographs in special copper ruby compositions by exposing the glass through conventional negatives and then heating the glass. And I had also made a new advance in scientific knowledge. These initial successes were exciting; I felt that a door was opening into the unknown, with endless possibilities beyond.

I was now becoming aware that glass at high temperatures is indeed a chemical solution, one whose ingredients can react with one another as in a water solution; but the enormous temperature range over which glass exists, together with the fantastic changes in viscosity from that of lubricating oil at the melting temperature to solid, brittle glass at room temperature, makes glass an unexplored world of its own. Sophisticated research on physical and optical properties of glass had been done, but very little had been done on high-temperature chemical reactions.

Photosensitive Gold Ruby

Now, I thought, the chemistry of the beautiful and rare gold ruby glasses probably resembles that of the copper ruby, so why not develop a photosensitive gold glass? The first experiments confirmed that indeed the same special reducing ingredients are required for a gold ruby, but the trick of changing these agents did not succeed in producing photographs. Where was the missing link?

At this point, I started a search of the literature on gold ruby glasses, which became a fascinating project in itself. The first report of a gold ruby was given by Antonio Neri in 1612, before the dawn of chemistry and in the age of alchemy. Neri, an Italian priest whose patron was a wealthy noble, wrote in his book, "L'Arte Vetreria," that he had made small precious stones of glass by repeatedly melting crushed glass with finely powdered gold, made by heating an aqua regia solution of gold. Neri's book was translated into English by Merret in 1672, and into German by Kunckel in 1679.[1]

Johann Kunckel (1630–1703) a famous alchemist and the descendant of several generations of glassmakers (his father was alchemist to the court of Holstein), invented the first practical formula for gold ruby glasses, building on the work of earlier alchemists, particularly Hollander and Schwartzer (as noted by Ganzenmúller). Kunckel's beliefs are a blend of medieval superstition and more modern understanding. For example, he rejected the idea of the alkahest (universal solvent), but he accepted the possibility of transmutation of metals and he believed in the supernatural curative powers of gold and gemstones (including synthetic gems made by recognized alchemists). According to Ganzenmúller, ". . . it was obvious for him (Kunckel) to think of preparing a genuine gold glass which, made into vessels, imparted to the liquid contained in them the virtues of the carbuncle (philosopher's stone). . . ."[2]

Ganzenmúller continues, "The impulse to the preparation of a true ruby glass was then given, as Kunckel himself tells us, by Dr. Cassius, who had suggested using, in the glass batch, his "purple of Cassius," a dye color made by reacting tin chloride with gold chloride. Simply, the key invention of Kunckel and Cassius was to introduce gold and tin compounds simultaneously into the glass batch or possibly into the molten glass. No real improvement in gold ruby glass, and no real understanding of the chemistry, would be forthcoming in the next three centuries.

Kunckel was first chemist and apothecary to the duke of Lauenburg and then to the Elector of Saxony, John George II, who put him in charge of the royal laboratory in Dresden. In 1679 he became director of the laboratory and glass works of Brandenburg. It was during his tenure here that he invented the gold ruby glass. It was an instant success and — since the Elector had a monopoly on it — it commanded a very high price. Of course, Kunckel also gained from his discovery, becoming famous and wealthy in his own right.

But—then as today—jealousy overcame judgment on the part of one of Kunckel's workers, one Christoph Grummet. Grummet's motives are not recorded, but the results of his perfidy are history: Grummet wrote a book in which he falsely accused Kunckel of using cheap saltpeter (potassium nitrate) instead of gold in the ruby glass. Well, that sounds reasonable. Who wouldn't use something cheaper than gold if it were possible? Today, yes. But in the medieval and renaissance periods, gold was the ultimate metal, not only for its beauty and value as a monetary unit, but, again, for its supposed ability to restore health to the ailing and retain it in the well.

Why would Grummet have done such a thing? It is well known that Kunckel was outspoken about the deceptions of fraudulent alchemists, and perhaps he was hitting too close to home for Grummet's comfort. In any case, Kunckel finally stopped making gold ruby because, as he said" . . . gold ruby has now become so cheap and common that it is for workers and not for great gentlemen."

And now the story becomes confused. According to Ganzenmúller— who was always willing to add a little drama to a tale —Kunckel lost his job, moved to Sweden, and lies in an unknown grave. That's a very touching tale. Unfortunately, it doesn't agree with the historical facts. The truth is that in 1688, Charles XI— who knew a good thing when he saw it— brought Kunckel to Stockholm, gave him the title of baron von Lowenstjern in 1693, and made him a member of the council of mines. He died on March 20, 1703 at Dreissighufen, his country house near Pernau, and there he is buried.

I resumed my experiments, feeling a bond across the centuries with the alchemists. Sure enough, gold dissolves as positive ions in oxidized glass and remains colorless, even when reheated. In a glass reduced with carbon, gold never dissolves because it remains in the metallic state; in a glass containing a tin compound, however, gold dissolves as ions and forms a gold ruby color on reheating because the stannous ions reduce gold ions to insoluble metal. But tin or no tin, exposure to ultraviolet light had absolutely no effect. I had not invented a photosensitive gold ruby. Now what had gone wrong?

Remembering from chemistry class and from photography books that a photochemical reaction requires that the effective light must first be absorbed by the medium, I searched for a sensitizer which could absorb ultraviolet light and then reduce colorless gold compounds dissolved in the glass to insoluble metal. Finally, one sensitizer, cerium oxide, was found, and another invention made, i.e., photosensitive gold glass, capable of photographs in blue, purple, or ruby. (The color varies with the size of the submicroscopic gold crystals, controlled by the photographic process of exposure and heat treatment.)

Dr. W.H. Armistead, my officemate and good friend (later to become Vice Chairman of the Board), then invented a photosensitive silver glass with yellow and brown coloration.

Photosensitive Opal Glasses

Next it occurred to me that my original objective, photosensitive opal glass, might be achieved indirectly. Here was the idea: By including the key sensitizers of photosensitive glass in the thermally opacifiable, sodium fluoride opal glass, we might first produce a photographic image of tiny crystals of gold (or copper or silver) and then the sodium fluoride opal crystals would deposit on, and only on, the metal crystals. This would develop a photosensitive opal pattern.

To make a long story short — I failed again! I was baffled, but not discouraged. Despite adding the photosensitive ingredients to thermally opacifiable sodium fluoride glasses, and decreasing the sodium fluoride concentration in small decrements until the glass remained transparent even on long reheating, the expected heterogeneous nucleation of sodium fluoride by metal particles did not occur. I couldn't understand it because I had faith in the principles of nucleation. (I had learned these principles in the thermodynamics course at M.I.T.)

I think it is only fair to the reader to describe the concept of "nucleation," technical though it may be. Nucleation is the unique key to every one of my inventions and, even more broadly, it is the initiator of every kind of change of state of matter in the universe. For example, when water freezes, the smallest stable particles of ice are nuclei, which grow to larger ice crystals. The ice nuclei can form spontaneously well below 0°C; but almost always they form on particles of dust in the air because this requires less energy. This is an example of heterogeneous nucleation, the dust particles being heterogeneous nuclei, or nuclei having a composition different from that of ice, the crystal phase being produced.

In glass, when a tiny crystal of gold, silver, copper, or other substance is present, it can become a nucleus like the dust in the air. This nucleus can initiate growth of other kinds of crystals in the glass at high temperature, as in the case of photosensitive opals and — as we see in later chapters — glass-ceramics of all kinds.

Backing off temporarily from sodium fluoride opals, I started exploring other glass systems. Within a short time, the lithium silicate system responded with a photosensitive pattern of gold-nucleated lithium metasilicate crystals, and the alkali baria silicate system produced a photosensitive barium disilicate opal glass. So, at least the general principle of heterogeneous nucleation (nucleation of one phase by a different one, as opposed to homogeneous or self-nucleation) was verified for these glasses. All of the three photosensitive metals, gold, silver, and copper, were able to nucleate these crystals.

Finally, the key to developing the photosensitive sodium fluoride opal was found. I found it one day by reheating the pre-exposed glass twice. Further trials showed that development requires growing metal crystals above 500°C, cooling below 400°C to nucleate, and finally growing nucleated sodium fluoride crystals above 500°C.* Beautiful three dimensional white opal patterns in clear glass could now be made in a photographic process. So my first major research goal had been reached and the exercise had taught me that glass, that cold and adamant material, has many secrets still to tell!

In due time, I prepared a technical paper presenting the new theory of the chemistry by which copper and gold produce ruby color and silver forms a yellow tint. After mailing it to the editor of the *Journal of the American Ceramic Society* (ACerS), I heard nothing for many months, then received a polite letter of rejection with no explanation of the reason it was turned down. Knowing that scientific journals rarely reject a manuscript without a reason, I wrote asking for an

* Dr. Robert Maurer of Corning later carried out brilliant research elucidating the nucleation kinetics, including proof that the metal particles must grow larger than 8 nanometers (80 A) in diameter to be capable of nucleating other crystals. The techniques that he learned in these studies of light absorption and light scattering of glass were to help him later in his landmark invention of the superpure silica glass waveguide fibers that now replace copper wire in telephone communications.

explanation. I learned that two of the three reviewers had recommended the paper but the third, the distinguished glass scientist, Professor Woldemar Weyl, had bitterly attacked it. The paper, "Coloration of Glass by Copper, Silver, and Gold," was eventually published by ACerS. It turned out that Professor Weyl was in the process of publishing his book, *Colored Glasses*, and my explanation ran counter to his. We later became friends, but neither of us changed his mind.

In 1950, I was granted my first 12 patents, some of which had been applied for as early as 1943. These patents included the photosensitive metal-colored glasses, the photosensitive opals, and some other opal glass patents, including thermometer opals (see Chapter V).

Although the invention of photosensitive sodium fluoride glass, FOTALITE®, was a rewarding technical success, and the products with delicate patterns of white crystals in clear glass could be beautiful, finding an appropriate market turned out to be a problem. It seemed (and still seems) to me that decorative tableware would be a natural field for this unique material, and we proved, with the help of Ray Voss and his people the General Development Department, that continuous production of tableware, chill-tempered for strength, was feasible, but the Consumer Products Division was not interested.

The management of the Lighting Products Division saw a potential market in 2 × 4 ft. lighting panels with a three-dimensional screen pattern of opaque grids to control the light: also, the possibility of architectural panels or windows was suggested by Tom G. O'Leary. Tom succeeded in selling this idea to Harrison and Abramowitz, the architects who were designing the United Nations Assembly Building. (These windows in translucent marble patterns are now part of the north wall of the United Nations Building in New York City and part of Corning Glass Center.)

As a result, full-scale production melts of rolled sheet glass were scheduled to be made in a factory at Kingsport, Tennessee. It so happened that only six weeks before the big million-dollar production trial was to begin, I made an important discovery. The nucleating agent for sodium fluoride had been a trace of gold (about 50 parts gold per million of glass). Suddenly I learned that silver could successfully replace gold as the nucleating agent, and that the use of silver decreased the heat treatment required to develop the crystallized pattern by a fantastic 100°C. This could greatly ease production problems because it would mean that problems of sagging, distortion, and warp would be eliminated. It seemed like a godsend for the new photosensitive opal glass. But so far, it had worked only for a few one-pound crucible melts of glass. Was it a fluke! Could it be repeated? Could it be extrapolated to a continuous production tank holding many tons of glass? My success record had not been so great as to inspire undue optimism, and my respected boss, Mr. Hood, went so far as to say he couldn't guarantee how secure my job would be if I made this gamble and lost! There were enough uncertainties already.

After considerable soul searching, I decided that it would not be fair to myself nor to the Company to settle for the safer path.

The gamble was made and won! We had a succession of successful production runs of the silver-nucleated FOTALITE®!

Polychromatic Photosensitive Glass: Full-Color Photographs

I suppose every long-time researcher files away in his memory or notebook a number of ideas or chance observations that constitute unfinished business — something to try later on when there is more time. This is the story of one such observation that lay dormant for 30 years, then flowered into an exciting invention.

Back in the 1940s, when I had been experimenting with silver nucleation of photosensitive sodium fluoride opals (FOTALITE®), I had learned that the silver requirement could be decreased if small quantities of reducing agents such as starch, sugar, or carbon were present in the batch. Surprisingly, one or two pot melts of tableware were found to exhibit a pale rainbow of colors after exposure and double heat treatment. All of the colors of the spectrum could be seen. Attempts to duplicate the effect failed, and further work was shelved as unfinished business.

In 1974, at a Consumer Products idea meeting. I made one "last" try to sell photosensitive opal tableware. Some interest was expressed and remembering the rainbow effect, I decided to try to understand and improve on it to extend the color range of photosensitive opal tableware. In the intervening years I had learned some interesting facts about silver colors in glass.

Dr. Armistead had produced glasses that he had aptly termed rainbow glass because they developed transparent rainbow colors when heated in a temperature gradient. Having read that aligned metal wires in glass result in a light-polarizing medium, I experimented with precipitating colloidal silver in glass and stretching the glass at high viscosity to elongate and align the silver particles. This had indeed resulted in polarizing effects. But additionally — and again to my surprise — red, blue, and green colors resulted, rather than the yellow of the normal glass that contained spherical particles of silver. In this case at least, the different colors must be due to different elongations of the silver particles. But must they be aligned?

With the theoretical advice of Dr. E.U. Condon, Dr. Roger Araujo and I had written a technical paper quantitatively describing the color changes of elongated colloidal silver particles, based on optical theory of metals and the adaptation of more general theoretical equations of earlier writers. The theory accurately described the color effects and predicted that color intensities could be extremely high. It seemed hardly possible that the photosensitive opal rainbow colors could be due to elongated silver particles since no stretching force had been applied to the glass. What could be the reason for those tints?

With regard to the earlier glass compositions, it soon became apparent that bromide additions intensified the colors, and that a second exposure between the first and second heat treatments greatly improved color. Furthermore — a surprise bonus — it became possible to develop transparent colors, all of the colors of the visible spectrum, as well as opaque colors. We had finally invented a full-color photosensitive glass, capable (with complex processing) of color photography. But we still could not understand the reason for the range of beautiful tints.

The first electron micrograph, made for me by Gerald Carrier, dramatically explained the reason: The silver was shown to be at the tip of a long, thin pyramidal crystal, and its shape was indeed elongated! Further electron micrograph studies correlated colors with particle shape and fitted very well with the theoretically derived equations. Improvements on the theory were made by Corning scientists Nick Borrelli and Jan Chodak. George Beall, Joe Pierson, and I coauthored a technical paper.

Could such a complex glass as polychromatic glass be made into sheet from a continuous melting tank? It appeared no more difficult than producing the complicated and volatile photochromic glasses (see below) that were by now in regular production. My chief concern was its great sensitivity to oxidation-reduction conditions which are controlled by batch ingredients in crucible melts and by tank atmospheres in continuous tanks. Also, glass technology people were nervous that devitrification might occur during the sheet draw, using the "overflow pipe" process.

A trial continuous tank melt was made at the Corning-Sovirel Avon laboratory in France, located near Fountainebleau. This was an interesting venture, not only in itself, but also because Ruth and I lived in a company-owned apartment in Avon for six weeks. The tank run successfully produced polychromatic glass sheet, thanks to the dedicated 24-hour-a-day labor of our French associates, Messrs. Thibieroz, JeanMarie, and A. Andrieu, as well as of Corning staff George Beall, Joe Ference, and George Luers. In gratitude to our French friends, we had a champagne party at the lab, and Ruth and I gave a farewell dinner at a downtown restaurant. We ate, drank, and sang until midnight. We have fond memories of Avon and its people.

In the belief that polychromatic glass can have eventual commercial success as a medium for color photography, Joe Pierson and I conducted experiments that proved the glasses are capable of producing colors whose purity almost, but not quite, matches that of commercial color transparencies. Whether polychromatic glass succeeds commercially or not, I am very proud of it as a scientific success and a new versatile art medium.

Photochromic Glass: Reversibly Darkening Glass

One day in 1959, Dr. William H. Armistead, then Vice President and Director of Research, called me into his office for a talk about an exciting idea.

"Don," he said, "do you remember the time I made those white silver chloride opal glasses that changed color in the light and faded in the dark?"

I did indeed remember them vividly. When Bill had been my office and labmate in our early years at Corning, he had patented photosensitive silver yellow glasses and also had produced silver halide opals with unusual behavior. But these impermanent photo effects had not appeared to be useful.

"I've been at a meeting with one of our ophthalmic glass customers," Bill told me, "and someone was saying how valuable it would be to have spectacles that would turn dark in sunlight, but clear again indoors. On the train coming home, I remembered my silver chloride opals and also remembered how you had made glass-ceramics[t] that are crystalline but transparent. How would you like to try to make transparent glass containing silver chloride crystals that darkens and clears reversibly?"

Being a confirmed optimist, I jumped at the chance. Perhaps no one but an optimist would have really tried, knowing the odds against success. After all, transparent glass-ceramics had been achieved by making the refractive index of the crystals equal to that of the glass. Since the refractive index of silver chloride crystals is much higher than that of any glass, any crystals would scatter light and cause opacity, unless the crystals are infinitesimal in size. Also, no perfectly reversible chemical reaction in a solid had ever been discovered, and that would be required if this invention were to be a success. However, by this time my faith in "miracles" had been strengthened by past experience, and this was certainly a worthwhile objective, since success would fill a need of humanity and find a ready market.

In my experimental approach, I started with some of Bill's silver chloride opal glass compositions and progressively decreased the silver chloride concentration until the glass, after forming and annealing, was transparent. The glasses, however, would not darken when exposed to ultraviolet or visible light. The transparent glasses were then reheated to below their softening temperatures since all of my past experience had shown the potency of such processing to nucleate and grow crystals or droplets of supersaturated ingredients in the glass. This treatment turned some of the glasses opaque, others translucent, and left still others transparent, showing that at least some precipitation had resulted. Some of the translucent and opaque glasses darkened in ultraviolet light, but disappointingly few, and some partially faded in the dark. None of the transparent glasses darkened. Had we reached the proverbial stone wall?

Having read the scientific literature related to silver halide photography, and also having seen how an optical sensitizer (cerium ions) had successfully photoreduced gold in my earlier photosensitive glass, I now tried adding possible sensitizers in trace quantities. Copper compounds proved effective. Before long, I had produced the first transparent glass that darkened and cleared.

But this was not yet success! The question still remained whether the reaction was really reversible or would the darkening and fading gradually decrease until the glass finally remained either dark or clear? No one would want spectacles that would wear out in a few hours, days, or months. Very fortunately, continuing tests proved that the glass never tires. Photochromic spectacles have become a successful product.

The chemistry of photochromic glass involves self-nucleated precipitation of supersaturated microdroplets of silver chloride (doped with copper ions) while the

[t] See Chapter IV.

glass is being heated at about 600°C, above the melting point of silver-halide crystals. Photo-darkening is analogous to the formation of the latent image in a silver halide photographic emulsion. Reversible clearing is due to the return of photoreaction products to their original sites in the microcrystals embedded in impermeable glass. Dr. Araujo and his colleagues, by their high-caliber theoretical and experimental studies, have advanced the detailed understanding of this reaction to a high degree of perfection and have improved the performance and composition dramatically. Index-corrected photochromic spectacles (eye glasses and sunglasses) in a variety of colors and forms are dispensed worldwide in multi-million-dollar quantities. The basic Armistead-Stookey patent was applied for in 1962 and issued in 1965.

Dr. Gail P. Smith, an excellent lecturer, writer, scientist, and Corning's former Director of Technical Services, tells the following story about the first use of a photochromic substance. During one of the campaigns of Alexander the Great, an old man came to him and showed him a cloth which, when dipped into a special dye, would change color with the changes in sunlight intensity. Alexander had a strip of this cloth issued to each soldier to aid in timing the start of an attack. This magic cloth became known as "Alexander's Ragtime Band!" (One magazine printed this as a serious fact.)

Some Bizarre Happenings with Photosensitive Glasses

Through the years since the invention of the first photosensitive glass, many unusual ideas for their application have been tried. Some included decorative art and tableware. Shortly after the invention of photosensitive gold glass, we were loaned the services of a Steuben "shop"— a gaffer, John Jansen, and his helpers — and a 1300-pound glass melting pot in "B" factory where Steuben glass was then made. Three melts of photosensitive gold glass were made and successfully formed into all sorts of goblets, vases, tumblers, plates, and bowls. Johnny really enjoyed the chance to "play" with a glass that was different and was glad to try special experiments during lunch time. Sometimes the glassworkers made beautiful paperweights on their own time. We decided to make a paperweight with Ruth's picture inside it on a disk of photosensitive glass. So I exposed the picture and gave it to Johnny, who carefully sealed it to a piece of Steuben glass and immersed it into a pot of molten red-hot Steuben glass to cover it with a dome of clear glass. During this operation, the photosensitive glass would receive plenty of heat to develop the picture.

But when the glass cooled, there was no picture! Red-faced, I went back to the lab. Theorizing that the fast plunge into hot molten glass must have destroyed the latent image, I duplicated this quick heating by dropping a pre-exposed disk into a hot oven at 800°C. Sure enough, no picture developed.

So I took another piece to Johnny and told him I had solved the problem. He had enough faith in me to decorate the paperweight with pieces of Millefiori (thousand flowers) cane. This time Ruth's picture appeared as if by magic in the finished paperweight, which we still possess. After exposure, I had slowly heated the disk long enough to start developing the picture.

This incident initiated research and scientific papers concerning the nature of the latent image produced by ultraviolet light in glass. When I asked Johnny to make a special piece of art glass for himself in which I would develop any design he wished, he made a beer stein that eventually had his own portrait in the bottom!

Glass Coins

During the Second World War, when a shortage of silver and copper was foreseen, the Secretary of the Treasury was looking for ways to replace the metals in coins. Dr. William Shaver enlisted the Secretary's interest in photosensitive glass. We made sample coins, first in copper red, later in white photosensitive opal glass. But cheaper metal coins won out, partly because their magnetic properties allowed them to be used in slot machines!

Cosmic Ray Research

Several thick bundles of photosensitive glass plates, carefully shielded from light, were sent into the stratosphere with balloons, as part of university research on cosmic rays. It was hoped to record tracks in a three-dimensional photosensitive medium. (I don't believe these experiments succeeded.) Exposure to a Cyclotron beam developed the ruby color without heat treatment!

Photosensitive Glass for the Office of Strategic Services (OSS)

When the patent application for photosensitive gold glass was filed in Washington, its issue was delayed for two or three years in accordance with government secrecy acts. The reason was that the glass could be used for carrying secret messages; that is, exposure to sunlight or ultraviolet light through a stencil or a photographic negative imprints an invisible latent image which can at any later time be developed to a visible image when the glass is heated to 600°C. Sure enough, we received a visit from two members of the OSS, and an order for all sorts of small glass articles that might be carried by a tourist: mirrors, spectacles, vials, etc. Dr. Erickson made the glass articles from an optical melting tank. I have never heard that a second order was received from the OSS and have often wondered what adventures the glass may have had.

FOTOFORM® Glass: Glass That Can be Photochemically Sculptured and Etched

"We must find an economical way to drill a million holes in a plate of glass!" This was the dramatic challenge given to half a dozen of us researchers one day in

1947 or 1948 by Dr. J. T. Littleton, our research director. It was in the early days of color television, and Corning, which has always been the major supplier of glass components for television sets, believed that the vital aperture masks guiding the electron beams should be made of glass. The only idea brought out at the meeting was to seal together a million small glass tubes, then saw slices from the bundle. This didn't sound practical to anyone.

Although I didn't say anything at the meeting, I went home wondering whether any of my photosensitive opal glasses could do the job. Unlikely as it seemed, perhaps some chemical or physical difference existed between the glass and the photographically crystallized areas that could be used to produce holes. Up to this point I had been blind to the possibility that photosensitive glasses might develop photographic patterns that differed in any property other than optical appearance. Dr. Littleton's challenge had startled me into a new concept.

Looking first for physical differences, I found that there were distinct differences in density and in thermal expansion. However, these differences were not great enough to produce the controlled microcracking that might cause holes to form. The next step was to learn whether chemical differences could be found. Differential solubility in the crystallized photographic pattern, on immersion in some magic solvent, was what I hoped to find.

After preparing several pieces of each of my photosensitive opals (sodium fluoride, barium disilicate, and lithium metasilicate) with a square screen pattern of transparent and opaque sections, I immersed each in a different chemical solution: strong acids, strong alkalis, and hydrofluoric acid, respectively (hydrofluoric acid is a good solvent for silicate glasses).

Imagine my delight when the glass with the barium disilicate in hydrofluoric acid solution etched slightly more than the opal; the sodium fluoride glass had grooves at the edges of the pattern, and the lithium silicate glass contained square holes where the crystallized pattern had dissolved completely! "A miracle," I thought to myself. Another secret door had opened.

Needless to say, the work to make this "photochemically machineable" FOTOFORM® glass had only begun. Nevertheless, an exciting and useful new material had been invented.

In a pilot plant project led by Dr. J. H. Munier, Corning actually did produce prototype glass color television aperture masks having the required number and geometry of holes, but the industry had adopted metal masks instead. Another technical success but commercial failure (so far at least) was the making of high-quality, photoengraved glass, master printing plates by Dr. Harry Kiehl and Corning staff.

FOTOFORM® glass has taken almost 30 years to become a big business in its own right; it is now used in complexly shaped structures for electronics, communications, and other industries. Its invention also became a key event in the continuing discovery of new glass technology, proving that photochemical reactions, which precipitate mere traces (less than 100 parts per million) of gold or silver, can nucleate crystallization, which results in major changes in the chemical behavior of the glass. I must admit, however, that even this dramatic discovery did not open my mind as it should have to the powers and broad possibilities of nucleation in glass. It took a lucky accident to teach me this lesson.

The Invention of FOTOCERAM®: The Lucky Accident that Launched Glass-Ceramics

"Damn it, I've ruined a furnace!" This was my first thought when I saw the recorded temperature reading — 900°C!

The laboratory oven, with an automatic temperature controller, was being used to heat-treat a plate of pre-exposed FOTOFORM® glass at 600°C for some of our etch-rate studies, and the controller had accidentally stuck in the "on" position. Knowing that FOTOFORM® glass melts and flows below 700°C, I was certain that a pool of liquid glass had flowed onto the floor of the oven. Imagine my astonishment on opening the door to see an undeformed, opaque solid plate! Snatching a pair of tongs, I immediately pulled the plate out of the hot furnace, but it slipped from the tongs and fell onto the tile-covered concrete floor, clanging like a piece of steel but remaining unbroken! It took no great imagination to realize that this piece of FOTOFORM® was not glass, but something new and different. It must have crystallized so completely that it could not flow, even though the temperature was more than 200°C above the softening temperature of the glass. And obviously it was much stronger than ordinary glass.

It now occurred to me that the photonucleated glass, although not containing enough silver (the nucleating agent) to cover the head of a pin, had nonetheless been caused to crystallize almost completely to a finegrained ceramic object! The properties of such a material would not be those of glass, but would depend on those of the crystals. This accident had resulted in the first-known piece of glass-ceramic.‡ It led me to realize an important general fact about glass: All *glasses, and not only the FOTOFORM® composition, can theoretically be converted to crystalline bodies having new properties that depend on the nature of the particular crystals.*

Doors were opening wider, and now I could see glittering vistas in all directions! Why did such an important discovery occur so late in the 3500-year history of glass, and why was an accident necessary to bring it about? The answer to this question is humbling to me personally, and perhaps it will cause other scientists to ponder.

In fact, this discovery should have been made many years earlier, by any scientist with a general knowledge of chemistry and thermodynamics. It was well known that all glasses are supercooled liquids frozen in an unstable state and that

‡ Glass-ceramics - a family of new polycrystalline materials made from glass by a process of controlled nucleation and crystallization. Applications include opaque and transparent cookware, (CORNINGWARE® cookware and VISION®, respectively), nose caps for guided supersonic missiles, high-strength tableware, electrical capacitors and machinable ceramics, among others.

given a chance, all glasses will alter to the lower energy state of being crystalline. J. Willard Gibbs had already spelled out the thermodynamics of homogeneous and heterogeneous nucleation to alter unstable or metastable states of matter to stable equilibrium, so a wise scientist could have predicted glass-ceramics before the fact.

Perfecting of the FOTOFORM® glass composition, the photographic process, and the etching or chemical machining process, required a great deal of coordinated research. The practical studies were accompanied by theoretical and experimental studies which, in my opinion, have significantly advanced scientific knowledge in the field of high-temperature inorganic chemistry. The patents on the FOTOFORM® glass compositions had been applied for in 1946 and were granted in 1950. The process patent, "Sculpturing Glass," was applied for in 1951 and issued in 1953.

The FOTOCERAM® Story

When the properties of FOTOCERAM® (the high-temperature crystallized FOTOFORM®) were measured, we learned that this fine-grained crystalline material was truly much harder and stronger than ordinary glass and high in electrical resistivity. Many of the commercial products now being made are first chemically machined to shape a FOTOFORM® glass, then re-exposed and reheated to 850°–900°C to convert them to crystalline FOTOCERAM®. Lithium metasilicate is the silver-nucleated crystal present at 600°C that permits etching, while lithium disilicate and quartz are formed at higher temperatures (800°–900°C).

One day in 1956, Dr. William Shaver surprised me with a happy report that FOTOCERAM® had passed the test for "supersonic rain erosion!" He then explained that the Navy was searching for a nonmetallic material that could be the protective nose cone for supersonic missiles. One of the many stringent requirements was the ability to survive a rainstorm at supersonic speeds. Dr. Shaver had entered FOTOCERAM® in a test that included all kinds of glass and ceramics, and FOTOCERAM® had sustained the least damage! This encouraged us to believe that the new glass-ceramics could indeed have unusual and useful characteristics.

FOTOCERAM® did not have other properties essential for supersonic radomes (namely, thermal shock resistance and high transmission for radar tracking beams), but I was just then making the first cordierite[§] glass-ceramic samples, and I thought, if we were lucky again, that one of these samples might have all the necessary properties.

§ Cordierite: a magnesium oxide, aluminum oxide, silica crystal; 2Al2O3•2MgO•5SiO2

IV

The First Glass-Ceramics: A New Family of Ceramics Made From Glass

Given the general principle that all glasses can be controllably crystallized by the appropriate nucleating agents, the question arose: Out of the infinite world of glass compositions, the thousands of kinds of crystals that might theoretically be precipitated, and the smaller number of potential nucleating agents, where should we start?

Two important goals became apparent. One of these was to obtain patent protection by applying the principle as broadly as possible and as fast as possible. The other was to try to produce useful new materials and products. Suddenly, the research project changed from a low-key, oneman exploration to a recognized breakthrough, and every available glass composition researcher was enlisted to contribute his effort, particularly on the first goal. Each researcher studied an assigned number of ternary (three-oxide) glass composition systems, testing the potential nucleating agents.

I chose for myself two aluminosilicate systems: magnesia and lithia. The magnesia-alumino-silica system contains several kinds of crystals known to form useful ceramic bodies, these crystals being cordierite, forsterite, spinel, and quartz. The system is alkali-free, making it useful in electrical insulating applications.

The lithia-alumina-silica system was known to precipitate at least two crystal phases, beta spodumene and beta eucryptite, that had the valuable property of near-zero coefficients of thermal expansion.* Corning has been and still is very successful with low-expansion PYREX® chemical ware and ovenware, but a material with superior resistance to thermal and mechanical shock could be useful in more severe environments.

Selecting a composition near the ternary eutectic or lowest-melting composition in each system, I wrote batch compositions with different potential nucleating agents and had the glasses melted, cast into flat squares, cooled, and then slowly reheated until they crystallized or remelted.

How were the nucleating agents chosen? Approximately a dozen chemical elements or compounds are capable of spontaneously precipitating in colloidal dispersion from a homogeneous glassmelt as it cools or reheats. If the atomic structure of such a colloidal crystal is close enough to that of one of the crystal phases that can be formed, the energy barrier preventing crystal nucleation is lowered, and crystallization of that phase occurs. This reasoning had been verified by the similar

* Thermal expansion: Most substances expand on heating and contract on cooling. Brittle substances such as glass and ceramics break from the local stresses of rapid heating or cooling, except for those having low coefficients of expansion (small dimensional change on heating).

atomic structures of the nuclei (gold, silver, copper) and the nucleated crystals (sodium fluoride, lithium metasilicate, barium disilicate) in photosensitive opals.

Again to my happy surprise, Mother Nature was generous to the optimistic explorer. Metal crystals were not successful nucleators in the aluminosilicate glasses, but titanium oxide was far more successful than we could predict. Its use resulted in nucleation of not only many of the crystal phases known from the equilibrium phase diagrams, but also of many additional crystal phases, some of them known as "metastable" or nonequilibrium phases. We found valuable bonus properties in some of these metastable crystals. Only later did careful detective work by many scientists, using sophisticated electron microscopy, X-ray diffraction, and other techniques, prove how bountiful nature had been. It was learned that titanium oxide could precipitate as colloidal crystals, but that in these aluminosilicate glasses its primary effect is to cause the glass to emulsify as it cools. The cold glass, still transparent, actually contains billions of microdroplets of one glass in a matrix of a glass of different composition. This greatly amplifies the number of feasible nucleation possibilities and ensures extremely small crystals in the resulting glass-ceramic. It also alerts the explorer to more mysteries within the glass.

Cordierite Radome Glass-Ceramic

The first few melts of titania-nucleated glasses in the slilica-alumina-magnesia system resulted in good-looking glass-ceramics containing cordierite as the major crystal phase. After tests verified that strength, hardness, low expansion coefficient, and dielectric properties were promising for the proposed radome application, samples were given to the Navy labs for their test program. In a remarkably short time, Corning was awarded a feasibility contract for manufacture and testing of radomes.

Now in a fast-moving development and testing program, newly invented glass melting and glass-forming equipment, together with specially designed heating equipment, precision finishing machines, and sophisticated instruments for precision measurement of radar frequency transparency, were all put together by Corning's mechanical genius, Jim Giffen. Code 9606 glass-ceramic, our first true glass-ceramic, is still being used by the Navy and Air Force for supersonic missile radomes.

As another unexpected bonus, I found in subsequent heat-treatment studies of Code 9606, the first of a new series of superstrength glass-ceramics — the subject of other chapters and inventions.

CORNINGWARE® Cookware from Low-Expansion Spodumene Glass-Ceramics

Exploration of the lithia-alumina-silica system nucleated by titanium oxide was found to produce the predicted crystals of beta spodumene or beta eucryptite, known to have very low, and in the case of beta eucryptite actually negative, coefficients of thermal expansion. Conventional ceramics had never been produced with these crystals because they expand along one axis while shrinking along another, which cracks the ware, but I optimistically believed that the minute crystals characteristic of glass-ceramics might not stress and crack the dish.

Small rods and flat plates had been measured, and had demonstrated very low thermal expansion coefficients but actual ware was not behaving well, to say the least. The heat-treated pots and pans all drooped like Dali watches, and at the same time were full of cracks.

Further experimentation finally resulted in the composition, Code 9608, from which CORNINGWARE® is made. The final composition contains some magnesia in addition to the other ingredients; beta spodumene is the primary crystal constituent. Except for having exceptionally high melting and working temperatures — even higher than PYREX®, Code 7740, Corning's highest-melting production glass, — Code 9608 had all of the properties required for nearly ideal cookware.

At this time a dynamic marketing genius, Lee Waterman, came to Corning. When Lee, the new General Manager of the Consumer Products Division, heard about the new thermal-shock-resistant glass-ceramics, he came to see us in the laboratory, inspiring us with his enthusiastic predictions of success in the market. He had been looking for something like this. As a result of Waterman's urgent push, a crash program was carried out, with Ray Voss playing a key role in pilot production and testing, and Dr. George Bair supervising pilot tank melting and forming. Everything went well, except for the extremely short life of the metal molds used for pressing the ware. (They melted!)

In spite of this problem, Corning's famous W.C. Decker, best known for his leadership in making Corning the television glass bulb manufacturer for the world and President of the Corporation, gave the go-ahead for a million-dollar production trial. Fast work by metallurgists and engineers solved the mold-life problem.

Lee Waterman's leadership carried CORNINGWARE® smoothly into the huge success he had predicted, and the same composition is still being made after more than 25 years, although many attempts have been made to improve it. Mr. Waterman became Vice President and General Manager of Consumer Products, and later President of Corning.

I'm sure that none of the people who were on the battle line during the development phase remember the project as being easy. In fact, when I expressed worry that our business competitors would get into the CORNINGWARE® business, George Bair said, "If they did, it would serve them damn well right!" Another memorable statement was made by Lee Waterman. He said, "If the patent department can give us a three-year lead in the marketplace, no one will ever catch us!" His prediction proved true, in spite of the fact that three major competitors did try hard. Anchor Hocking's effort resulted in a landmark court trial, a patent infringement case brought by Corning, described later. My broad "glass-ceramics" patent was applied for in 1956 and issued in 1960. *Annual sales of CORNINGWARE® cookware have been in eight figures for many years.*

Meanwhile, on May 23, 1957 (coincidentally, my 42nd birthday), Corning dedicated its new Houghton Park complex and laboratory buildings. The national press was invited. The first public announcement of PYROCERAM® was a highlight. The whole affair resulted in lots of publicity for the company and for me personally. I remember both my saying to Dr. Eugene C. Sullivan (Corning's premier research director, retired at the time) that this was the high point of my life and his flattering reply that it would not be the last one.

Incidentally, sad but impressive proof of CORNINGWARE® cookware's heat resistance exists in photographs of homes that have burned to the ground leaving only the CORNINGWARE® undamaged!

Transparent Low-Expansion Glass-Ceramics

During my exploration of the low-expansion, lithia-alumina-silica glass-ceramics, the method of screening the various glasses was to have a cane or rod hand-drawn from each one-pound crucible melt. These were cut into four-inch lengths and given a range of heat treatments, then examined for appearance and routinely sent to the physical measurements lab to determine their expansion coefficients and densities. The untreated transparent glass was also measured for comparison. If the heat treatment or the nucleation were insufficient, no crystallization nor change in appearance or properties occurred. As expected, crystallization was accompanied by alteration to white opacity, with a major decrease in expansion.

One day when comparing the data for a series of rods, I observed that some rods remained transparent after heat treatment, but noted that expansion coefficients had decreased to nearly zero.† The first time this happened, I assumed that a mistake had been made. After the second time, I carefully repeated every step. Lo and behold it was true: Some rods remained perfectly transparent but unquestionably had crystallized to zero-expansion ceramics! Another "lucky" accident, but a good invention would have been lost if careful observation had not been combined with optimism.

This phenomenon (crystallization with retention of transparency) was one of my biggest surprises. Later sophisticated studies by Dr. George Beall and others revealed relatively broad composition fields in which infinitesimally small crystals of solid solutions having a "stuffed beta quartz" structure were precipitated. These crystals are in a metastable phase, nucleated with the aid of "liquid-liquid" emulsification initiated by titania, and form transparent products.

I was reasonably, but not completely, successful in producing experimental transparent dishes having the thermal shock resistance of CORNINGWARE®. The pitchers had a slight brownish tint and some translucence, so that coffee in the pitcher looked suspiciously like motor oil!

---

† When the thermal expansion coefficient equals zero, the substance neither swells nor shrinks with temperature change; therefore, it is not stressed by thermal shock.

Dr. Beall later replaced titanium oxide with zirconium oxide as the nucleating agent, and produced truly transparent, colorless, zero-expansion pitchers and dishes. To our disappointment, the Consumer Products experts did not believe transparent CORNINGWARE® cookware would be saleable.

In 1980, Corning's French associates revived this invention after market tests showed that European housewives (who never bought the opaque CORNINGWARE® cookware in any volume) would like to buy this transparent form with an amber tint for top-of-stove cooking. Dr. Kenneth Chyung has further improved the composition by adding the bonus of an economical short heat treatment in the production process, and VISION® transparent cookware has become very popular in the European market. Subsequently, the transparent product was introduced into the United States and is still a major cookware product trademarked VISIONS. In 1998, CORNING divested itself of the Consumer Products Division so it could concentrate on the fast growing fiber-optics business.

Thanks again to Ray Voss and his hardworking helpers, CORNINGWARE ®, has been given an electrically conductive coating that makes it a leading utensil for microwave cooking. As another improvement, a special aluminum coating has been applied to its bottom to give it high thermal conductivity, enabling the cook to use it on top of the stove. These products, because of their added value to the cook, are also commercially successful. All of these products, together with PYREX® ware and CORELLE® tableware, have made Corning the leader in the houseware market.

CORNINGWARE® Patent Infringement Trial: Corning Glass Works vs. Anchor Hocking

Surprisingly soon, Anchor Hocking put on the market a copy of CORNINGWARE®, which Corning felt was a direct infringement of my broad glass-ceramic patent, U.S. Patent No. 2 920 971, issued January 12, 1960; Corning decided to sue for infringement. The infringed patent was in essence a broad umbrella covering the whole technology of glass-ceramics. It comprised 18 pages, 101 examples, and 21 claims. This trial became the most elaborate and expensive in Corning's history up to that time, involved many dramatic ramifications, and resulted in a landmark court decision important to high-technology inventors.

Clarence Patty, head of Corning's Patent Department, headed up Corning's legal team, with the very able assistance of Clinton Janes, a fine patent attorney who had written all of my patents since the earlier days of Walter Rising. Mr. Patty had retained W. Philip Churchill, a top patent trial attorney, partner in the New York firm of Fish, Richardson and Neave, who actually directed the strategy and prosecuted the trial in court. He was assisted by two young lawyers, Albert Fey and Lars Kulleseid.

As inventor, I was chosen to be Corning's principal witness, although we had several other important witnesses as I will relate.

My first task was to give to the opposition copies of every notebook and report I had ever written relating to the invention. It was somewhat of a shock to learn that such a legal requirement exists. It seems to me that an unscrupulous corporation could use it to learn the secrets of more advanced competitors.

After each side had reviewed the other's documents, several days of pretrial "discovery" took place. The opposition lawyer quizzed me and others about all the documents, digging for anything that might help their side. I did my testifying wearing a cast on the leg I had broken just a week earlier skiing. All this was recorded as sworn legal testimony and used later in court. Our lawyers did the same to the opposition witnesses.

Both sides engaged expert witnesses from outside both corporations. Two scientists from A.D. Little Corporation, D. William Lee and Paul E. Doherty, were engaged to do independent research related to our patent claims as to nucleation and crystallinity and report on them in court. Mr. A. Robertson, a representative of the Applied Physics Laboratory at Johns Hopkins, voluntarily testified for Corning with respect to the value of PYROCERAM® radomes to U.S. defense. Anchor Hocking's expert witnesses were Professor Rustum Roy of Pennsylvania State University, and geologist Dr. Hatten Yoder of the U.S. Geological Survey.

Phil Churchill went over the patent word by word with me, first to acquaint himself with the technology and then to discuss points that might be questioned by the opposition lawyers. At the same time, he searched for any trace of dishonesty in obtaining the patent, stressing the point that any such flaw could be a fatal weakness in our defense of the patent. Throughout the trial, our knowledge that we had nothing to hide strengthened us immeasurably as our testimony was mercilessly attacked by the opposition lawyers.

The trial was held in the District Court for the District of Delaware in Wilmington, Delaware under Judge Caleb Wright. A Wilmington attorney, Clair John Killoran, was our local attorney of record. The trial lasted 15 days.

I developed great respect and almost an affection for Judge Wright as the trial went on. He was not a scientist; in order to understand the facts he could have hired his own scientific expert, but he decided that I could serve the function. He became so interested in the facts that at one point he forgot himself and answered the lawyer's question before I could, whereupon he blushed bright red, and Mr. Churchill had the record erased.

Our side's strategy was first to explain the invention thoroughly and to prove its originality and usefulness, and then to prove that Anchor Hocking's cookware, and even its packaging and advertising, infringed my patent and constituted unfair competition. Anchor Hocking's strategy, since (as their documents indicated) they realized they were infringing my patent, was to try to show the patent was not valid, or that it had been obtained by fraud.

Our side presented its case first. I was on the witness stand for five days; three days answering Churchill's questions to explain the invention and two days being cross-examined by the defendant's lawyer, Arthur G. Connolly. His cross examination was the worst ordeal of my life. The only thing comparable was my oral doctoral dissertation examination at M.I.T., but this was longer and conducted by an enemy, trying to catch me in a mistake or a lie. Early in his cross examination, he referred to a PYROCERAM® radome as "an upside down spittoon." This backfired in two ways: First, Judge Wright stood up infuriated, and made a speech assuring that witnesses would not be insulted or intimidated; second, I decided this man could not and would not outwit me. Our witnesses proved beyond doubt that Anchor Hocking's cookware was well within the claims of the patent and really did infringe.

Anchor Hocking now began its defense case. No employee of Anchor was put on the stand. The first witness was Dr. Yoder, who, as it turned out, had done experiments on basalt. He testified that his results had come out the opposite of those reported by Corning's Dr. Herbert Kivlighn, thereby implying that Corning had committed fraud when prosecuting its patent application. (Kivlighn's work had been carried out in answer to a patent examiner's question about alleged similarity between basalt (frozen volcanic lava) and the products of my invention.) It was later shown that Dr. Yoder had reproduced the experiments, based on the patent information, and submitted them as surprise information.

Dr. Beall starred at this point, his geological expertise helping Churchill cross-examine Dr. Yoder so well that the latter developed real respect for Churchill's knowledge and toned down his assertions.

Yoder's experiments were surprise introductions of material to the trial, and Judge Wright would not have had to permit their introduction but he ruled that he would withhold his decision as to their admissibility. Meanwhile, Mr. Churchill demanded to see the documents describing the experiments.

In an all-night study, Clint Janes made a dramatic discovery. Dr. Yoder had not used the same basalt as Dr. Kivlighn, as he had testified under oath that he had. A chartered plane for a quick round trip to Rochester (Ward's Scientific Supply House), brought back proof that Janes was correct. When this new evidence was shown in court the next day, the effect of Dr. Yoder's testimony was demolished.

The testimony of Anchor Hocking's other witness, Dr. Roy, was directed not toward fraud, but toward showing that the patent was not valid because its claims were not sufficiently definite.

Specifically, he attacked the important claims describing the products of the invention as more than 50% crystalline. Speaking as an expert on quantitative X-ray diffraction, he testified that most people would not be able to measure it accurately. But Churchill, during cross examination, challenged him. "Could you, yourself, measure it accurately by X-ray diffraction?" Dr. Roy replied, "Yes, I could determine it within plus-or-minus eight percent!" We were elated at this answer, since it appeared that he had rebutted his own testimony.

This brief outline of the trial omits enormous contributions of high-caliber, painstaking, and time-consuming research by Corning's Gerald Carrier, proving that quantity and type of crystal can be measured by X-ray diffraction or with the electron microscope; by James Farricomb and his staff in determining these things by wet chemistry techniques; and by George Beall, whose tutoring of Mr. Churchill and whose own testimony played a great part in counteracting the attacks of Anchor Hocking's experts; by the Arthur D. Little men in isolating the crystal nuclei and determining their composition; and by Dr. Armistead and Mr. Lee Waterman, buttressing Corning's case.

When I left with my family for a European vacation, during the last day of the trial, we all believed we had won. To our astonishment and dismay, Judge Wright's verdict was in favor of Anchor Hocking!

Although most of his decision was as favorable to Corning as though our own lawyers had written it, even stating that ". . . the invention disclosed in the patent is a basic and pioneering advance in glass and ceramic technology," his final conclusion was: "However, the patent is unenforceable because, in 1956 when the application was made and in 1960 when the patent was issued, to determine the percent crystallinity within this 7 to 10 percent margin of error required costly and lengthy independent experimentation to devise a test to ascertain whether a product was within the bounds of the patent claims."

In effect, Judge Wright was writing a new patent law, harmful to complex high-technology inventions today and tomorrow, saying that complicated inventions are not patentable.

Corning appealed the decision to the United States Court of Appeals for the Third Circuit in Philadelphia, Pennsylvania. This court reversed Judge Wright's decision and ruled in favor of Corning holding that "(i) in our view, the District Court erred in finding Corning's patent to be unenforceable since the standards placed upon this particular patent were, as a matter of law, too strict." The court emphasized Judge Wright's finding that the invention was ". . . a basic and pioneering advance in glass and ceramic technology" and then declared, "Where the court is confronted with such a pioneer patent, liberality becomes the keynote of construction requiring the court to give the patentee a wide breadth of protection in construing the patent claims and specification."

That last sentence is quoted in textbooks on patent law. The court concluded that "from the facts and circumstances concerning this invention and under the sound law of the case, the only reasonable conclusion is that the patent was sufficiently described so that those who are skilled in the art were apprised of the patent claims and specifications without being forced to engage in costly independent experimentation."

So the trial came to a happy ending for our side — except for the enormous cost to Corning in money and in man-hours of technical and legal (patent attorneys') time, as well as in lost research time. Much of the research done in support of the case, however, resulted in significant advances in quantitative knowledge of nucleation and crystallization.

V

Those Devilish Opal Glasses

The various white opal glasses are among the most difficult to manufacture because the growth of crystals that opacify the glass occurs while the molten glass is cooling and being pressed, blown, or otherwise formed into a product. As one might suspect, any difference in cooling or in thickness of the ware may spoil the piece's uniformity of appearance. Thus, it is almost impossible to devise small, inexpensive lab experiments to model the exact results to be expected when tons of ware are being made per day. The production run itself becomes the final expensive test.

The Tricky Thermometer Opal

Not long after I had begun research into opal glasses, an urgent problem developed in the factory. For many years, Corning had manufactured most of the white-backed glass tubing with accurate capillary bore from which mercury thermometers were (and are) made. (The white back behind the mercury column helps one to see the column.) In fact, the 150-foot thermometer updraw tower, painted white, with the blue figure of a glassblower near the top, was at that time the tallest man-made structure in the town.

The thermometer tubing had always been made by a "handshop," composed of the gaffer, the gatherer, and their helpers. A huge gob of clear lead glass would be built up on the end of a big blow-iron and shaped into a tapered cylinder with a lengthwise hole. Next, a smaller gather of glass would be made from a clay pot of molten opal glass and carefully smoothed onto one side of the original hot cylinder. Then the cylinder would again be immersed in the original clear glass pot and covered with a final layer of clear glass. After final marvering (rolling on a steel table) to reshape it, the opposite end of the cylinder was sealed to another iron tool, the purity iron. In the older method of tube drawing, the men of the shop would draw the tubing horizontally, walking one end away from the other and fanning the hot spots to obtain uniform dimensions until the tube cooled. Then it was laid on the floor, cut into four-foot lengths, and annealed to remove the stresses. The tower updraw method later improved the geometry of the tubing by stretching the hot, viscous cylinder vertically using a mechanical pulley system.

Now, modern technology was taking over the production of thermometer tubing: ingenious engineers had designed melting and drawing units for continuous mechanical production, eliminating the handshop crew.

Briefly, pieces of the white opal glass are remelted in a vertical U-shaped refractory pipe with a small orifice situated inside the larger tank of clear molten lead glass; the molten glass is then drawn upward with the clear glass so as to form the white-backed thermometer tubing in a continuous mechanical updraw operation. The operation was a mechanical success, but otherwise a failure. The several problems that prevented making a satisfactory product were all traced to the opal glass. It was learned that the opal glass that had produced good hand-drawn tubing was not good for the machine-drawn product. Opacity was insufficient, and the tubing developed a mysterious tendency to warp in the hands of the thermometer fabricators, Corning's customers.

After some cogitation, I concluded that the problems arose from the radically different thermal history associated with the new process. The opal glass, homogeneous and clear at its melting temperature in the pot, had been given ample opportunity to crystallize and opacify during the several slow coolings and reheatings given it by the handshops. On the other hand, the single, fast cooling cycle associated with the machine drawing had not permitted the complete crystallization reaction to occur. This would explain both the insufficient opacity and the warping because reheating during subsequent flameworking to fabricate the thermometer would cause the opal glass to crystallize further, changing its density, and setting up unsymmetrical stresses that would warp the tube.

My first attempt to solve these problems was to employ a thermally opacifiable opal that would remain transparent throughout the drawing process and could be opacified by a heat treatment only slightly above the annealing range* of the glass. This did improve opacity, but I was very disappointed to see that every tube bent in an arc during the reheating that developed opacification. It finally became apparent that crystallization in the opal glass was still changing its density and its expansivity; since the opal glass formed a stripe along only one side of the clear glass tube, these changes resulted in warp.

I then concluded that we really needed something new: a glass that would melt to a clear homogeneous glass at 1300°–1400°C, cool to a clear glass, but opacify in the remelt unit. In other words, this glass would opacify completely while being reheated, just prior to being drawn into tubing. Thus it would be opaque and also completely reacted so that no warp would occur on later flameworking or annealing.

This theoretical solution left several questions. What crystal phase would fulfill both the requirements and the stringent condition that the opacifying crystals would not grow into large "stones" while being held in the hot box for many hours? I feared this could create the glassmaker's gremlin, devitrification; as the opacified glass remained in the "white box" for hours or longer, large crystals might grow and break the tubing. After searching the literature and consulting my colleagues for all of the possible opacifying phases, I learned from Mr. Hood about cadmium sulfide and selenide ruby and yellow glasses; sometimes the colloidal cadmium sulfide grew larger crystals, resulting in opaque yellow glasses.

I experimented with chemically reduced glass (as opposed to oxidized glasses which would form sulfates) and learned that thermally opacifiable white zinc

* Annealing range: temperature at which glass is held to remove internal stresses.

sulfide opal glasses could be successfully used in the new machine production of thermometer tubing. The ZnS crystals formed a fine-grained opal at 900°–1000°C, above the drawing temperature of the tubing. The new product was made by CORNING for many years, until the business was sold to another company. It is the subject of one of my first patents, applied for in 1945 and issued in 1950. I believe this type of dense opal glass still has possibilities for other applications such as tableware.

This experience taught me that an industrial researcher must bring together the many strings of a complex problem to bring it to a conclusion, to my mind, a more difficult and rewarding task than that of the academic researcher who studies one variable of an artificial system.

The Wellsboro Opal Light Bulb Adventure

One great difficulty with research and development in the glass industry is that the introduction of a new glass into commercial production requires a million-dollar gamble. The melting tank and its controls and accessories, and the machinery for forming ware continuously day and night, plus the skilled engineering and labor crews, constitute a major investment for the corporation. It therefore behooves the researcher to be sure he has done his preliminary work as well as humanly possible before taking the final plunge; but yet, once he believes he is right, to be brave enough to take that plunge. The following adventure will illustrate these points, and perhaps teach some lessons to the young glass researcher.

Ever since Edison's first incandescent glass light bulb, efforts have continued to increase lighting efficiency. One direction of research has been to produce white light-diffusing bulbs to hide the filament and spread light uniformly. This can be done by coatings or by using an opal glass.

As with thermometers, handshops were used to make good opal light bulbs. However, demand greatly exceeded capacity, and the Wellsboro ribbon machines were continually spewing out clear bulbs at phenomenal rates, so it seemed a fine opportunity to do the same with opal bulbs.

The only major glass composition problem, it appeared, was to have a glass that was clear and homogeneous on melting, which would opacify fast enough during forming to serve the purpose. Attempting to simulate the production method on a laboratory scale, we found that a fluoride opal with lots of fluoride would satisfy this requirement. Production-scale pot melts with bulbs formed by handshops seemed to confirm the conclusion. Finally, a full-scale production trial was run in the Wellsboro, Pennsylvania factory, producing opal bulbs with the ribbon machine.

Opal bulbs were made, but the experiment failed, for two unforeseen reasons. First, the bulbs had a spotty appearance because the machine-made bulbs were not absolutely constant in wall thickness in different parts of the bulb, which showed up as differences in brightness. Second, the water used to cool the molds reacted with the hot glass to produce hydrofluoric acid vapors that even frosted factory windows. Hindsight says these problems could have been foreseen, but they weren't; so far, they remain unsolved.

The Charleroi Barium Phosphate Opal Dinnerware Trial

Since the Middle Ages, bone ash (calcined animal bones) or calcium phosphate has been an opacifier both in fine china and in glass. Its use in glass, however, has been limited because it is not very soluble in molten glass, tends to form large crystals at high temperatures, and is not very opaque.

Hoping to find better behavior with other phosphates, I experimented extensively with barium phosphate and with lead phosphates. These behaved much better, although they too had their problems. For one thing, a microscope furnace study revealed that it was difficult to determine their liquidus temperatures, and it finally turned out that, above the melting temperature of opacifying crystals, the glasses go through a temperature zone of emulsification, i.e., a two-liquid phase separation. This means that, during cooling from the melting temperature of 1400°C or higher, an emulsion liquidus temperature is reached (say at 1165°C) below which a barium-phosphate-rich molten salt is insoluble and forms an emulsion of tiny liquid droplets of molten salt. If cooled relatively fast, these droplets remain tiny, crystallize at a lower "crystal liquidus" temperature, and thereby produce a beautiful white opal glass.

A tricky aspect of these glasses, though, is that if the temperature is held for appreciable times just below the emulsification liquidus, the molten salt droplets continue to grow and coalesce and, being heavier than the glass phase, finally sink to the bottom of the vessel to form a layer of molten salt. A glass was finally developed which, handworked from laboratory crucibles, with pot melts and day tank melts of up to 5 tons of glass, made beautiful opaque lustrous dinnerware. The chief concern was whether opacity would "strike in" rapidly and uniformly enough when the ware was pressed by an automatic forming machine from a continuous tank orifice from which the clear glass emerges at high temperature.

Finally, the crucial production trial was made, in a production continuous tank at Charleroi, Pennsylvania, where most of Corning's opal tableware was produced, and expert melting and forming "shops" were available. A fatal mistake somewhere in the melting instructions resulted in the tank temperature being held for two or three days at the worst possible temperature—just below the emulsion liquidus! As a result, when the glass was presumably melted and ready to form, lo and behold, it was solid as a rock and would not flow! Worse yet, when my friend Andy Wagner, Charleroi's Chief Glass Technologist, tried to drain the tank while standing below it in a pit and removing a bottom plug, a stream of white-hot molten salt squirted down, exploding as it struck the cooling water in the pit. I don't know how Andy escaped alive and unhurt, and have always felt guilty about my not having expected this consequence, and thereby endangering Andy's life!

VI

Changing the Nature of Glass

The remarkable changes that are brought about when glasses are nucleated — changes in physical, chemical, optical, and electrical properties — led me to take a broader look at the practical opportunities for improving glass in the direction of structural applications. By this time, scientists all over the world, as well as a fine group at Corning, were also developing new glass-ceramics after learning of our discoveries. Now British, German, French, Russian, Japanese, Czech, Romanian, Chinese, and other scientists were all in the game. I was pleased to learn that even the Russian literature had credited me with the basic invention.

It may seem strange to think of glass — a weak, brittle, but hard-to-machine substance — as a load-bearing structural material. Remember, however, that windows are important parts of most buildings. Glass is fire-proof, resistant to corrosion, and lasts forever if not broken. Perhaps it can be made really indestructible. Perhaps it can be made tougher; more readily drilled, sawed, and machined. This kind of thinking initiated many new experiments and discoveries.

Unbreakable Glass and Glass-Ceramics

Chill-tempering (well illustrated by Prince Rupert drops*) was in active use at Corning before my arrival, chiefly for strengthening PYREX® opal glass tableware used by U.S. Defense Forces, hotels, and restaurants. This quenching treatment produced a surface compression layer such that even abraded glass was twice as strong as annealed ware.

How could glass be made even stronger? It could evidently be still stronger if the compressive stress in a thin surface layer were still further increased. Harrison Hood called my attention to still another facet of glass science: ion exchange.† It occurred to us that ion exchange could alter the chemical composition of the surface layer of glass, possibly providing a method for chemically developing the desired surface compression. Mr. Hood called to my attention an old Corning patent by Edward Leibig that had described high-strength glasses produced by copper staining of glass.

This sparked a period of ion-exchange research in which we learned that glasses could indeed be chemically strengthened to about 25,000 pounds per square inch by ion exchanging smaller ions, such as lithium ions, for larger ions in the

* Prince Rupert drops are beads of glass with long tails made by dropping molten glass from the end of a rod into cold water. They cannot be broken with a hammer but shatter into powder if the tail is broken.

† If a molten salt containing sodium or potassium ions contacts a hot glass containing potassium or sodium ions, ions can be interchanged on a one-for-one basis.

glass, such as sodium or potassium, at a temperature above the annealing temperature. Cooling such glass articles resulted in the desired compressive layers. This invention was the subject of a patent by Hood and Stookey issued in 1957.

A still better invention was later made by Dr. Kistler. It consisted of reversing the ion exchange, introducing larger ions for smaller, at temperatures below the annealing range. This produced still higher compressive stresses, up to 60,000 psi. A shot in the arm was now given to the research program when Mr. W.C. Decker let it be known that top management were interested. In meetings led by Dr. Armistead, the various efforts on the subject were coordinated and before long Drs. Martin Nordberg, Ellen Mochel, and Joseph Olcott had perfected glasses strengthened by ion exchange. Thin sheets of the glass, trademarked CHEMCOR©, were very flexible. A daring incursion into the automotive windshield market, based on the proven superior safety of the new glass, was spearheaded by Dr. Armistead. It nearly succeeded and was headed off only by quick action of the major glass corporations to improve their own windshields. This was an impressive step in the direction of unbreakable glass, and probably approached the limit of practical tensile stresses attainable in conventional glass.

The advent of glass-ceramics initiated a whole series of further steps in strengths beyond those possible with glass. Knowing that some kinds of crystals (beta spodumene, beta eucryptite, and beta quartz) are characterized by zero and even negative expansion coefficients, my colleagues and I experimented with surface-nucleated crystallization of these types of crystals on a glass article. Cooling such an article from above the annealing temperature of the glass causes the interior to shrink, leaving the surface constant in volume or even expanded, so great compressive stress can result in the surface.

This work resulted in compressive stresses of 80,000 to 100,000 psi, and led to strengths at least ten-fold that of annealed glass. In order to demonstrate that this kind of strength approaches "unbreakability," Dr. Olcott and I made transparent glass cups having a negative-expansion eucryptite surface layer and dropped them from the roof of Corning's nine-story office building onto steel plates laid on the ground. These cups bounded several feet into the air, but did not break. (A movie of this exists somewhere.) Dr. Harmon Garfinkel and Daphne Rothermel also strengthened glasses by ion exchange, which resulted in surface crystallization of low-expansion crystals.

Meanwhile, during exhaustive heat-treatment studies of the cordierite radome glass-ceramics Code 9606, I found astonishing strength, together with higher-than-usual expansion coefficient, when the "ceramming" temperature was lowered to 1000°C from the usual 1200°C. Strengths were up to 150,000 psi even after abrasion. This work was followed up by Ray Voss, and he invented tamper-proof "frangible" products that are very difficult to break, and, if broken, have so much fracture energy that electrical alarms can be triggered. It was discovered that these products developed a surface layer of one type of crystals on a glass-ceramic core of another type of crystals, the skin having lower density and lower expansion coefficient than the core, such that the high relative compression resulted in the composite glass-ceramic article.

Still more recently, Dr. David Duke and Dr. George Beall developed composite glass-ceramics in the nephelme system (sodium, aluminum silicate) with strengths of 250,000 psi. These are truly, for all practical purposes, unbreakable! These superstrength glass-ceramics owe their strength to a high-temperature ion exchange within the crystals themselves, which substitutes larger potassium for smaller sodium ions and expands the crystalline surface on a crystalline core. Beall and Pierson have again found a remarkable bonus effect—namely, that when objects of these enormous strengths are purposely cut open, a deep saw cut is required before they will break spontaneously. In other words, they are not sensitive to surface flaws, as is glass. Perhaps still more nearly unbreakable glass-ceramics will be developed. But I believe that we already have materials that can be lightweight structural products for the future.

Undersea Strength: Compressive Strength

Still another aspect of super strength in glass and glass-ceramics stems from the nearly infinite compressive strength of these materials. When glass breaks, it is always from tensile stress, and the superstrength materials we have discussed up to now are also high in tensile strength. But submarine devices are subject to ever-greater compressive stresses at increasing depths below the ocean surface, stresses that exceed the strength of concrete, aluminum, and even steel.

I'm sure that others first conceived of deep-sea glass submarines; but at least I had a part, with Bill Baldwin and Dr. Shaver, in initiating undersea compressive strength tests of glass-ceramic (Code 9606) torpedo shapes and large PYREX® glass spheres run by the Naval Research Laboratory with R. Perry as consultant. These tests confirmed that glass and glass-ceramics have excellent strength-weight values in deep-sea conditions and can even survive explosions. I believe this is another future frontier for super-strength glass structural materals.

Machinable Glass-Ceramics

Hoping to invent softer glass-ceramics that could be sawed and drilled, I remembered soft minerals such as soapstone, talc, and mica. These minerals contain water in the crystals and, therefore, are not stable at glass melting temperatures. It was also known that water can be replaced by fluorine in these compounds, so I began experiments with melts of appropriate compositions and invented a forerunner to mica-type glass-ceramics in 1963. The patent was issued in 1967. George Beall, using sophisticated crystallographic know-how, has perfected them to a commercial product (MACOR®) which is finding increased markets each year. This mica glass-ceramic is especially useful because it combines machinability with strength and excellent dielectric properties. Since then, hydrosilicate glasses have been made that are both transparent and machinable, and we have also made mica hydro-ceramics by way of the autoclave process. This is another field for further development in inventing structural materials of the future.

The Search for Toughness

Although great strides have been made toward making "unbreakable" glass and glass-ceramics by increasing tensile strengths to 60,000 and 250,000 psi, respectively, these substances still remain brittle. That is, they flow or yield very little when stressed. Consequently, they may not be ideal structural materials because they cannot be readily sawed and drilled and any break is instantaneous and complete. The machinable micaceous glass-ceramics can be readily sawed and drilled, but they are still somewhat brittle and do not so far have super strength.‡

We believe there is still room for development of new structural materials combining strength and toughness so that once a crack has been started at a surface, the material resists continuation of the break through the interior. Many tough materials, such as metals and bamboo, consist of two phases: a hard, strong, stiff phase to carry the load, and a rubbery matrix to dissipate the stress.

Jadeite Stone: Weapons of New Zealand Maoris

On a vacation trip to the South Pacific, I became intrigued in New Zealand by museum specimens of polished translucent green stone axe heads, spear points, and other weapons and tools made and used by the Maoris, the warlike Polynesians who crossed thousands of miles of ocean in seven canoes to first populate the two islands. The knowledgeable museum director told us the stones were a form of jadeite, a very tough material, highly prized. Tribal chiefs made their own weapons with meticulous care, rubbing the edges on their stomachs for the final sharpening. When the British later "discovered" and invaded the islands, the Maoris never surrendered, and the two races have always had equal rights.

I brought samples of the jadeite home with me for analysis and crystallographic studies, hoping to synthesize a glass-ceramic like it. The stones did have a strange microstructure that resembled intertwined curving threads. So far, we have not duplicated this structure exactly, but Dr. Beall has made tough glass-ceramics with other long-chain crystals. Using his education as a geophysicist, Dr. Beall has explored many silicate-chain glass-ceramic possibilities, and has developed products that are indeed much tougher than any previous glass-ceramic. Some of these are low in cost, easily manufactured, and already earmarked for significant markets.

This exciting research comes at a time when technological development in many fields demands high-performance properties not met by known metals or plastics. All sorts of new engineering and structural products made from glass and glass-ceramics will be produced in the coming decades. The rapid progress made in our generation toward producing materials and products that are virtually

‡ Tensile strength is about 25,000 psi.

impervious to all destructive forces (fire, mechanical stress, chemical attack, etc.) leads me to encourage the next generation of scientists and engineers to learn how to employ indestructible structural products to man's benefit and to learn to avoid unhappy consequences.

I believe there are strong arguments for this approach. For one thing, a look to man's near future shows the disappearance of raw materials for metals, petrochemical plastics, and similar structural materials, and wood has always been an impermanent and dangerous substance for construction. The raw materials for glass, glass-ceramics, and ceramics will always be available and cheap. For another thing, the terrible costs of deterioration of structures by weathering, rust, fire, wind, and water are incalculably huge and can be reckoned not only in money, but in human lives as well.

Indestructible objects have problems of their own, but they can be "designed around." If we assume, for example, that a structural panel or beam of indestructible material "X" once fabricated cannot be altered, we must design it so it can be used without subsequent machining, bending, or other special handling. Even machinability and field fabrication can be combined with virtual indestructibility in the new inorganic "ribbon" board and paper being made from reconstituted glassceramics by Beall, Hoda, and their colleagues.

The Search for Rubbery Glass: Hydrosilicate Glass

One of Ambassador Amory Houghton's[§] continuing challenges to me was to invent a rubbery glass, preferably unbreakable. It is a remarkable fact that the strength of a perfect, flawless piece of glass is extremely high — almost a million pounds per square inch. Yet because of its brittle nature, glass can be easily scratched and thereby become weak. If the surface were rubbery, the glass would be strong because scratches would not propagate into the interior. The only solid clue to a rubbery silicate glass that I had ever found in the literature was a book entitled *Soluble Silicates* by Vail.[7] In it is a description of making "water glass" solution by dissolving sodium silicate glass. Vail had observed a residue in the bottom of the vessel, a hard, rubbery mass that was tough and resisted breaking by a hammer blow.

This statement was enough to start me on a new research trail. I began making rods of sodium silicate and potassium silicate glasses and exposing them to the steam of an autoclave. It soon became evident that soft or rubbery layers could be developed by this treatment and that they contained dissolved water. The steam reaction was unique; the same treatment in boiling water dissolved the glass, and in acid solution, leached out alkali, leaving porous glass. Rods or ribbons having the soft, reacted layers proved very strong (about 200,000 psi) because the steam reaction had removed surface flaws and formed a protective coat. But unless this coat had been sealed against water evaporation, the strength was temporary, diminishing as the coating hardened and cracked. I still believe applications of this

§ Former Corning Glass Works Board Chairman, former U.S. Ambassador to France

phenomenon are possible in laminates and/or in hydrated granules for making strong composites or rubbery products.

Another interesting invention based on steam treatment was "self-destructing" hydrosilicate glass containers. These containers were designed to serve the usual purposes, then, on removal of a tab when opened, to progressively disintegrate to powder as water evaporated from the surface. So far, their added manufacturing cost has kept them from the market. A combined group of dedicated workers from Technical Services, the General Products Development Department, and Research Department made this a technical success.¶ The hydrosilicate glasses behave in many ways like very low-melting glasses, and can be formed and molded in the same temperature range as some thermoplastics. This had led Dr. Helmuth Meissner into extensive research in the low-temperature forming area. He and Dr. Richard Maschmeyer and Dr. Roger Bartholomew have succeeded in producing finished aspheric lenses (requiring no further forming or polishing) by their new, low-temperature forming techniques.

Partly spurred by this work and partly by problems associated with the water in hydrosilicates, other researchers, including Dr. Paul Tick, have now developed new, durable, low-melting, water-free glasses that can also be molded into lenses by plastic forming techniques. These new glasses appear to be supplanting the hydrosilicates, even before the latter succeed, but at least the hydrosilicate research has spurred other marketable products.

Still another field of inventions — hydrosilicate ceramics, analogous to glass-ceramics — has opened up as a result of our hydrosilicate research. As we widened the search for glass compositions susceptible to steam treatment and increased autoclave temperatures and pressures, some of the glasses crystallized internally, some producing hydrated and others nonhydrated microcrystalline products. High-magnesia glasses produced machinable tetrasiliceous mica ceramics, for example. Our first patent of steam treatment was issued to Hugh Bickford, Loris Sawchuk, and myself in 1970, having been applied for in 1967. It has been followed by many others, with a number of Corning inventors represented.

An Unexpected Bonus: New Process for Porous Silica Glass and Low-Temperature Glassmaking

While we were experimenting with soluble sodium silicate glasses, I conceived the idea that they might react chemically with organic polymers to produce plastics combining the good properties of both. After some review of polymer chemistry, it appeared to me that ureaformaldehyde polymers might be suitable candidates. Some interesting experimental plastics resulted. But before long, Joe Pierson and I learned that simply heating appropriate proportions of paraformaldehyde with sodium silicate solution gradually formed a white solid. Further study showed that after washing and drying we had a solid piece of porous amorphous silica; that the pores were continuous, uniform and interconnected; and that the pore size could be

¶ The group included R. Bartholomew, D. Campbell, H. Dates, S. Lewek, F. Marusak, J. Pierson, B. Swinehart, and myself.

accurately predetermined by the proportions of ingredients. Dr. Helmuth Meissner carried the chemical studies further and made another invention, finding that other reagents that slowly released acid to the solution could produce similar products. A Pierson-Stookey patent and a Meissner-Stookey patent were issued in 1974. Since then, Dr. Robert Shoup has developed a low-temperature chemical process for making glass articles of pure silica and also other glass compositions. In the future, as energy sources diminish, this low-temperature, chemical, glassmaking process may become the way to manufacture many glasses.

Stookey prepares to expose a portrait to ultraviolet light (1950). Photo courtesy of Corning Incorporated Department of Archives and Records Management, Corning, NY.

Ben Allen, James Giffen, William Armistead, and S.D. Stookey with PYROCERAM® radomes. Photo courtesy of Corning Incorporated Department of Archives and Records Management, Corning, NY.

Stookey receiving the Sullivan Award from Eldon Sullivan. Photo courtesy of Corning Incorporated Department of Archives and Records Management, Corning, NY.

Experimental FOTALITE® dinner plate. Photo courtesy of Corning Incorporated Department of Archives and Records Management, Corning, NY.

FOTOFORM® etched sculpture, from a continuous-tone photograph. This scultpture demonstrates that depth of etch varies with exposure. Photo courtesy of Corning Incorporated Department of Archives and Records Management, Corning, NY.

Bridal photo of Ruth Stookey: a transparency in photosensitive gold glass. Photo courtesy of Corning Incorporated Department of Archives and Records Management, Corning, NY.

FOTOFORM® glass honeycomb with live honeybee. Bees filled such a honeycomb with honey. Photo courtesy of Corning Incorporated Department of Archives and Records Management, Corning, NY.

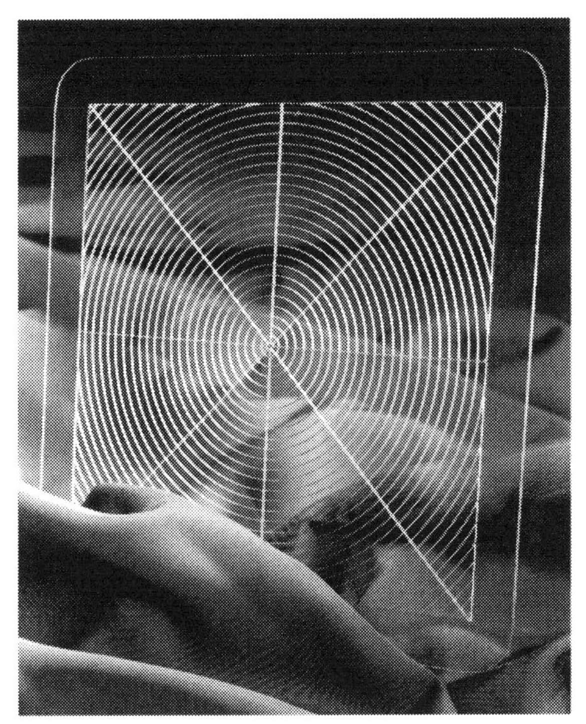

FOTOFORM® "spider web." The glass ribbons are not much thicker than a human hair. Photo courtesy of Corning Incorporated Department of Archives and Records Management, Corning, NY.

FOTOFORM® glass lace. Photo courtesy of Corning Incorporated Department of Archives and Records Management, Corning, NY.

Expressions dinnerware. Photo courtesy of Corning Incorporated Department of Archives and Records Management, Corning, NY.

Corning VISIONS® cookware. Photo courtesy of Corning Incorporated Department of Archives and Records Management, Corning, NY.

Stookey with a display of Corning Ware.

President Ronald Reagan presented the Nation Medal of Technology to Stookey in ceremonies at the White House in 1986.

In 1997 and 2000 Stookey visited Africa on photo safaries. In this 1997 photo, he shows off the day's catch, a Nile perch.

Stookey with a six-foot, 75-pound tarpon he caught and released in February 2000 in Key West, Florida.

VII

Reflections on the Science of Glass and on Pioneering Research

At school, I learned in the Introduction to Thermodynamics course that all matter at equilibrium is in one of three states: solid, liquid, or vapor. Where does glass fit? Actually, it is a liquid at high temperature, but a nonequilibrium solid on cooling. Because of its enormous increase in viscosity, it freezes all atomic translational motion while the solid is amorphous and never crystallizes to its equilibrium state, crystals being the state of lowest energy.

Glass, therefore, in its usually observed state resembles none of our usually recognized states of matter. To a chemist, its unique characteristic consists of the fact that it is a metastable* medium, with all of its chemical ingredients (even electrons, in some cases) frozen in nonequilibrium states. Superficially, the chemistry of glass might seem intractable and uninteresting, but as the reader has seen, this is untrue. Glass chemistry is high-temperature chemistry of oxide solutions, together with lower temperature viscosity-controlled reactions.

Molten glass at white heat (1400°–2000°C) is a true solution of oxides in melted silica sand. Pure silica glass, even at 2000°C (well above the melting temperature of sand crystals), is almost too stiff and viscous to form, but has so many properties, including fantastic transparency, that it finds important uses, such as the recent Corning-initiated use as optical waveguides for communications.

The hot glass is a nearly universal solvent capable of dissolving compounds of every chemical element. Its chemical composition, therefore, can be infinitely variable. Chemical reactions of all kinds can occur between the various dissolved species, some of which we have seen in this book. In the future, more chemists will be turning their attention to investigating such reactions. In predicting the direction of reactions in glass, the laws of thermodynamics are the controlling influence; but in determining the rate, or determining whether a reaction will even occur, the viscosity controls.

J. Willard Gibbs has emphasized the importance of the nucleation phenomenon in all transformations of matter. In most of our experience, phase changes or chemical reactions occur so rapidly that we are unconscious of the nucleation step. Seeding of the clouds is one familiar example. In solid glass (in contrast to a water solution, for example), reactions are so slow as to be nonexistent. This permits the ingenious scientist to introduce selected nuclei, in selected localities if he wishes, to guide subsequent reactions in desired directions. Then, by carefully reheating to

* Metastable states are those in which a substance is frozen in apparent equilibrium, but can change to a more stable lower energy state. Glass, for example, can be changed to more stable crystals.

temperatures (and lower viscosities) where electrons, atoms, and molecules can be selectively freed to diffuse and react, the desired reactions can be carried out, or stopped at any intermediate point by cooling the material.

Glass, therefore, can be the ideal medium for studying the initial kinetics of many kinds of chemical and physical reactions and crystallization phenomena. And the more we learn, the more valuable new inventions can be made. The possibilities are infinite, since every chemical element can be incorporated in glass. A whole world of hitherto unknown polycrystalline materials has already been found in the glass-ceramics so far developed, in addition to those predicted from phase equilibrium diagrams. They include many metastable crystal phases and solid solution crystals, some having valuable properties such as transparency, low expansion, and high strength. Various crystal conformations (plates, ribbons, threads, pyramids, etc.) have been tailor-made in glass-ceramics. Further variations may be unlimited.

When I reminisce about the first seven decades of my life, I feel that I am one of the fortunate few who has reached and even exceeded the high goals he had set in his youth and has enjoyed a well-rounded life as well. Like an explorer in a newly discovered land, I have pioneered in a new field of science, the chemistry of glass, and found previously unknown treasures both useful and beautiful. If this in itself were not sufficient, other tangible rewards have been many. They include a number of nationally recognized awards for scientific achievement, for significant invention, and for useful contributions to humanity. Not the least reward is in the knowledge that hundreds of researchers worldwide are even now carrying on where I left off, making more and more discoveries. One of the greatest rewards is the warmth that comes from recognition by my personal friends at Corning, including factory workers who stop me on the street to thank me for their jobs.

Pioneering can have its drawbacks, too, including loneliness and self-doubt. The following episode illustrates the first of these. When I was ready to present my discovery of photosensitive glasses to the scientific world, I was confronted by a dilemma. Who would know or care what I was saying? I was a member of both the American Chemical Society and the American Ceramic Society, but few chemists knew anything at all about glass, and the only Ceramic Society member who might be interested was bitterly hostile. So I submitted an abstract for an international Chemical Society symposium to be held in New York. Out of 10,000 papers given at the meeting, mine was the only one on the subject of glass. The program committee put me in a session on elastomers and plastomers, where I felt very lonely. The paper did attract the attention of the scientific news media, probably because of its novelty. Since that time it has become a habit to speak to people with various interests—the American Physical Society, the American Ceramic Society, the British Society of Glass Technology, American Optical Society, American Photographic Society, American Society of Mechanical Engineers, and others. The National Academy of Engineering has done me the honor of electing me to its distinguished membership, even though I am not an engineer. And when in 1971 the American Chemical Society gave me its National Award for Creative Invention, I felt like the lamb come back to the fold.

A Review of Successful Commercial Products Based on My Inventions

In many ways I have felt akin to the alchemists, artists, and sculptors of the Middle Ages, whose material welfare was taken care of by wealthy patrons so they could devote all their energies to their work. For almost 10 years, the products of my labors were mainly discoveries in the chemistry of glass and the photosensitive glasses which were regarded as interesting and beautiful, but not utilitarian or profitable. Finally, the President, W.C. Decker, became a little impatient and passed down the word that I should look for more practical things to do. Only recently, Dr. Armistead, then my officemate, was reminding me of the day that our Research Director, Dr. Littleton, came into our office and gave me that message, then told me to continue as I was doing, but to keep it *sub rosa* for awhile! Fortunately, the FOTOFORM® discovery, with its breakthrough possibilities in color television, came just in time to restore us to Mr. Decker's good graces. The chemically machinable glass took at least another decade to find its real niche, and is now used in a multitude of applications, in complex and miniature shapes for computers, electronic displays, electronic printers, and recently, as decorative collectibles.

The thermometer opal, specifically formulated to solve an already existing problem, found immediate commercial sale, and has been in continuous production from about 1950 to the present time.

The first glass-ceramic, formulated for the especially rigorous demands of supersonic guided missiles, also found immediate use and still has not been superseded, although speeds and other requirements have become still more severe. A basic reason for this is the perfect reliability and reproducibility of all properties both within a large radome and from one piece to the next. This is the advantage of glass-ceramics, made from perfectly homogeneous glass, over any other ceramic or composite structure.

CORNINGWARE® cookware, with its combination of resistance to breakage by extremes of heat and cold and its attractiveness, was on the market within a year of its discovery in 1957. It became and still is a mainstay of Corning's consumer product line. It has been claimed that enough CORNINGWARE® has been sold to provide every U.S. home with more than half a dozen pieces. When the white CORNINGWARE® cookware was introduced in Europe, it was not a success, and Corning learned that European cooks prefer to cook on top of the stove, where they can see and stir their concoctions. So, with improvements in the transparent low-expansion analog made by Dr. George Beall, Dr. Kenneth Chyung, and a French colleague, A. Andrieu, an amber-tinted, top-of-stove family of cookware was made and sold first in France about 1980, then in all of Europe, with great success. A cautious sales test has since been run in the United States, whose favorable outcome is encouraging a full-scale introduction of this new form of cookware in this country, trademarked VISIONS®.

Photochromic glass, the reversibly darkening glass which Dr. Armistead and I invented, was first marketed as prescription spectacles in the late 1960s. With periodic improvements in the range of colors and reaction rates, and with sunglass versions, these optical products comprise another multimillion-dollar business. Special types have been developed for sportsmen and, very happily, also for people afflicted with eye problems such as retinitis pigmentosa and cataracts. Recently a plastic competitor for photochromic spectacles has appeared on the market.

Dr. Beall's relatively new MACOR®, the softer glass-ceramic that can be sawed, drilled, and machined almost as easily as wood, is daily finding more uses as a do-it-yourself ceramic, and is one component in the exterior of the space shuttles.

CENTURA® is a thin, high-strength glass-ceramic dinnerware invented by J. F. MacDowell, which was popular for a number of years. It was then replaced by CORELLE®, somewhat similar in appearance, but less costly, made from two laminated glasses whose strength, good appearance, low weight, and low cost have made it the most popular tableware in America. CORELLE®, not my invention, is manufactured from flat sheet by a continuous process devised by an inventive engineer, James Giffen.

Except for FOTOFORM®, the photosensitive glasses have not yet found significant markets, but there are good indications that they will soon be prominent as energy-conserving and decorative windows, skylights and other architectural products.

Meanwhile Russia, Romania, and Czechoslovakia have developed glass-ceramics for panels, large piping, and paving; the British have their own formulas for electrical insulators; and the Japanese are very active in FOTOFORM® and transparent glass-ceramics. Even China is developing glass-ceramics and photochromic glasses.

I think it is safe to say, as others have already said, that my research has started a revolution in glass and ceramic technology and that the results will be long-lasting.

IX

Some Extracurricular Adventures

Dayspring — Housing for the Elderly

Ruth and I have been active members of the Methodist Church ever since her arrival in Corning. About 1970, I filled out a church questionnaire, saying I would be interested in projects to help elderly people. Soon I found myself a member of a small committee dedicated to build housing for senior citizens, and not long after that I became chairman. We searched a long time for a suitable site, and the disastrous flood of 1972 finally provided one for us in the downtown Urban Renewal Area of Corning. An expanded interdenominational commitee now applied to HUD, the government housing authority, and was granted a position as Preferred Developer, based on approval by local authorities. So I made several appeals before the Mayor and the City Council, who refused our application. We turned to a neighboring town, but then local opinion and newspaper editorials in our favor helped the Mayor change his mind. Now our city has two new apartment buildings in the Civic Center, attractively landscaped — home for 500 elderly people and open to all races and creeds. Entertained by church organizations and operating their own clubs, the inhabitants are a happy group. I resigned the chairmanship when the first building's success was assured and remain proud of my modest part in this civic contribution.

A College Professorship

Shortly after retiring from full-time research, I was honored by an invitation from the Dean of Alfred College of Ceramics to be a Visiting Professor for a semester, replacing a man who was taking a sabbatical leave. I'm not at ease as a lecturer, but I accepted the challenge, teaching a class of 30 seniors and graduate students a course on "Nucleation and Crystallization," a course already in the curriculum. I had great rapport with both students and faculty and was treated like royalty. The students apparently regarded it as a privilege to attend class and really absorbed the information. Toward the end of the course, I gave the class an assignment to "invent something." To my surprise and pleasure, nearly every one submitted original ideas, some so good that I hastily returned them because they may become patentable inventions! At a farewell dinner that Ruth and I gave to the class, they presented me with a fine plaque, inscribed "To Prof. Stookey, from the Best Class He Ever Taught."

Other Adventures

Lest the reader accuse me of being strictly an armchair adventurer, and believing the other sides of life to be also important, I will mention some of my extracurricular adventures.

When I was five years old, our family migrated from Nebraska to Iowa by way of New Orleans and Galveston, Texas. We camped in swamps, surrounded by alligators and cottonmouth moccasins; my parents were nearly lost in a hurricane while fishing off Galveston. And they nearly lost me on the bustling New Orleans docks when I dallied to watch South American monkeys being offloaded from a ship.

Later, we had many family fishing trips to Minnesota, tenting out among mosquitoes and other natural phenomena. I remember silently questioning why we went through so much misery to have fun, but my brother and I both developed a love for fishing which, I'm glad to say, has carried through to my sons and their sons. At home in Corning, New York, I enjoy stream fishing for bass and fly-fishing for trout, as well as taking youngsters to the ponds in May to catch panfish. I've caught marlin and sailfish off Mazatlan, Cozumel, and the Florida Keys, squaretail trout, with my son Don, in lakes in Quebec; survived a seaplane crash, with brother Dave, on a fishing expedition beyond the Arctic Circle on Great Bear Lake; eaten moose meat shot by our Indian guide on Makokibatan Lake in Northern Ontario, and later worn moccasins made by his wife from its hide. My brother, my older son Bob, and his two sons David and Steven, shared in this experience.

I'll describe some of my sensations during the seaplane wreck because it was one time in my life when I thought I was about to die, and it was one time that my brain sent me completely illogical messages.

We were landing in the lake near the base camp after a day of fishing at another lake. Our plane was a four-passenger Cessna with leaky pontoons and a huge, young pilot. Dave was in the copilot's seat, another fisherman and I in the rear. The lake was like a mirror. The pilot made the mistake of landing perpendicular to the wake of a large companion plane and we flipped ("water-looped") so that the tail of our plane was vertical, then tipped forward until we were upside down in the frigid water. The upside-down cabin filled with water. I was certain that the plane was diving to the bottom of the 1300-foot deep lake and that we would drown. No sensation of fear struck me at that point, but one of sorrow for Ruth and Don who were at home and might miss me.

Looking upwards, I could see the water surface and realized the plane was partially afloat. Although the plane was upside down and my safety belt and that of my seatmate had come unfastened, by some quirk of gravity we had remained right side up. From that moment until the next day, my confused brain told me the plane had landed right side up! Fortunately, we were able to open the door against the pressure of the water and we two rear passengers crawled out and stood on the upsidedown wing, hanging onto the floating pontoon with our heads above water.

I held the door open for what seemed an eternity waiting for my brother and the pilot to appear. Finally, I realized that they might be hurt or trapped and just when I was trying to summon the courage to dive back in, Dave's face appeared! Hanging upside-down under water, he had had trouble unfastening his belt and finding his way to the exit. Dave kept an illusion that he had been swimming under water out in the lake, until he had run out of breath. Meanwhile, the pilot had escaped by kicking out the windshield. A dozen boats raced from the lodge to rescue us, and we were soon shivering uncontrollably and drinking brandy.

Next day, watching fish swim inside the plane, I finally realized it was actually upside down. Dave and I took a 30-mile boat ride in rough waves. He caught a 28-pound lake trout and I found out that I had some painfully cracked ribs.

The third fisherman? When we left on the 1000-mile return south to Edmonton, he was refusing to ever step into an airplane again! He may still be there. Would I have gone back under water to rescue my brother? I hope so, but I'm very glad he made it by himself!

Ruth and I made two six-week driving tours of Europe, once with another couple (Howard and Jane Lillie) and later by ourselves. We've also visited the Caribbean Islands both by plane and by cruise line, with my mother and her brother, Randall. And we've toured the Orient by commercial plane, even as far as fabled Katmandu in Nepal near Mount Everest.

We've had a surplus of hair-raising boating adventures, with 28-foot Owens Motor Cruisers and a 36-foot Trojan, over 30 years, including cruises across Lake Ontario, on the St. Lawrence to the Montreal World's Fair, down the Inland Water Way to the Florida Keys, complete with dangerous storms on Chesapeake Bay that nearly capsized our 36-foot jinx cruiser, Carefree Lady, and ending in a collision with a reef whose warning buoy had shifted in a recent storm, wounding Ruth and grandson David. I sold the cruiser on the spot, for a third of its value.

The beautiful, crystal-clear Homosassa River on Florida's West Coast yields many happy memories of fishing with Mother, Dad, and various family members among scenes of tropical beauty.

The notorious Hurricane Agnes of 1972 flooded our Pritchard Avenue house to near the ceiling of the first floor, but did not affect our morale as badly as might have been expected. We were not in Corning at the time; instead, we were riding out the same hurricane, with 70-mile winds on our boat, Carefree Lady, in the dock at Cambridge, Maryland, off Chesapeake Bay. When we returned, the house had already been partly mucked out by Ed and Margaret Zak, our close friends, and our daughter Margaret's parents-in-law, and by our son Bob and his wife Sally Jo (nee Parsons).

In 1965, when I was 50 years old, I ran out of research ideas and into a period of depression, feeling that I had nothing more to contribute. At this one low time in my life, my boss made some personnel changes which, I wrongly thought, agreed with my self-assessment as a has-been, and I submitted my resignation to him.

He wouldn't hear of it. When I stubbornly insisted that I wouldn't stay on any longer, he sent me to talk to both Amory Houghton, Jr. and Sr. All three were more sympathetic and complimentary to me than I deserved, and the upshot was that instead of resigning or being fired, I was given a paid leave of absence for a year. Ruth and I traveled most of that year, over most of the world. Then in November, while Ruth stayed at home to get ready for Christmas, I shipped our 28-foot single engine Owens motor cabin cruiser to Annapolis and cruised down the Inland Water Way, including Chesapeake Bay, to the Florida Keys with my brother Dave and his wife Marie.

That year really refreshed me in body, mind, and spirit. I came back to Corning eager to work and full of ideas, and will always be grateful to those responsible for being so understanding and kind to me. (Jamie Houghton, then in Europe, and John Guth, steered us to some wonderful spots in southern France.)

My lovely wife Ruth and I were married for 53 years, until her fatal heart attack. We have had three children, each of whom married and had two children, all grown now, and I even have a precocious great-granddaughter who has a new little sister. We have all been close. The men and boys have been together for annual deer hunts and frequent fishing and camping trips to isolated fly-in lakes in northern Ontario.

Robert Alan, born in 1942, is now a Rochester, New York obstetrician-gynecologist with degrees in chemistry from Hamilton and in medicine from Columbia College of Physicians and Surgeons, having done internship and residency at Strong Memorial Hospital in Rochester. Bob married Sally Jo Parsons of Corning. Sally has a master's degree in special education from State University of New York at Buffalo. Their older son David is a XEROX researcher married to Phyllis, and they have two daughters. Steven, a college graduate, is a sales manager at PAYCHEX.

Margaret Anne was born in 1944, graduated from Bucknell, and married Corningite and co-Bucknellian Edward Zak, who is now a veterinarian in Virginia. Their son Ben married Jennifer Cooper, and they are in business near the Outer Banks of North Carolina, both college graduates. His younger brother Darren is completing his education.

Donald Bruce was born Christmas Day, 1949. Don graduated from Corning East High School, but spent his senior year and graduated at the American School in Lugano, Switzerland. He has a degree in government from Hamilton and a master's degree in social service from Syracuse, After a number of years as a student counselor in Utica schools, he is trying another vocation making use of his talents in handicrafts and art. Don is married to Elizabeth Nye of Rye, New York, a graduate of State University of New York at Utica-Rome, New York. She is a teacher of special education in their home town, Utica. They have a lovely daughter Melissa (the only girl of my six grandchildren), who graduated in three years at SUNY Buffalo and had a dangerous year teaching in Albania for the Peace Corps, before preparing to teach in North Carolina. Jarek (Jake) graduated at SUNY Binghamton, majoring in computers, and is now a computer technician in Vermont.

I know that my family has been my finest achievement.

Epilogue

Nanocrystals in Glass

On looking back at the scientific discoveries and inventions described in this book, it becomes apparent that many of the valuable properties of the materials (in addition to the characteristics unique to the crystal composition, such as low expansivity) can be attributed to one factor that all have in common: namely, the extremely small dimensions of the crystals or crystallites in their microstructure. The diameters of the particles range from those of single atoms to 1000 Angstroms.

Photosensitive Polychromatic Silver Glasses

These glasses are unique, in that—although the particles are too small to scatter light—their full spectrum of colors is due to a range of shapes. Electron micrography shows that the spherical particles have a sharp absorption peak in the blue, resulting in the pure yellow known in color photography; while progressive elongation results in two absorption peaks, one moving into the ultraviolet while the other moves toward longer wavelengths through the visible spectrum. Our research has shown that this behavior agrees quantitatively with theoretical equations based on the optical constants of silver (See references 25, 26, and 27).

Photochromic Glasses

These glasses, employed in spectacles because they darken reversibly and continually in sunlight, fall into the rare class of "smart" materials. Their chemistry is related to conventional photography, in that their response to light depends on doped silver halide crystals. Why is their behavior different? Obviously because of the infinitesimal size, combined with the impermeability of the glass matrix. The transpency of these glasses is also due to minute size. Larger silver halide particles cause opacity, because their refractive indices are much higher than that of glass.

Glass-ceramics

Glass-ceramics—made from amorphous glass but almost 100 percent composed of submicroscopic crystals—have physical properties characteristic of the crystals. Throughout history, the larger crystals accidentally formed by devitrification have been anathema to the glassmaker because they cause breakage due to expansivity differences. Some kinds of crystals, including those in CORNINGWARE and VISIONS (lithium aluminum silicates) have large differences in expansivity along different axes, so that conventional ceramicware containing such larger crystals would be weak. Here again the small size and flaw-free nature of the particles in CORNINGWARE has made possible the use of its thermal shock strength.

VISIONS cookware has the same crystals as CORNINGWARE, except that its crystals are too small to scatter light; hence its transparency (a mixed blessing, since it looks like ordinary glass!).

Addenda

Publications

1 . E. C. Bingham and S. D. Stookey, "Relation Between Fluidity, Temperature, and Chemical Constitution of Pure Liquids," J. Am. Chem. Soc., **61**, 1625-30 (1939).
2. S. D. Stookey, "Photosensitive Glass, a New Photographic Medium," Ind. Eng. Chem., **41**, 856-61 (1949).
3. S. D. Stookey, "Colorization of Glass by Gold, Silver, and Copper," J. Am. Ceram. Soc., **32**, 246-49 (1949).
4. S. D. Stookey, "Chemical Machining of Photosensitive Glass," Ind. Eng. Chem., **45**, 115-18 (1953).
5. S. D. Stookey, "Recent Developments in Radiation-sensitive Glasses," Ind. Eng. Chem., **46**, 174-6 (1954).
6. S. D. Stookey, "Nucleation," pp. 189-195 in Ceramic Fabrication Processes. Edited by W. D. Kingery. John Wiley & Sons, New York, 1957.
7. S. D. Stookey, "Catalyzed Crystallization of Glass in Theory and Practice," Ind. Eng. Chem., **51**, 805-808 (1959).
8. S. D. Stookey, "Glass-ceramics," Mech. Eng., **82** [10] 65-8 (1960).
9. S. D. Stookey, "Controlled Nucleation and Crystallization Lead to Versatile New Glass-ceramics," Chem. Eng. News, **39**, [25] 116-25 (1961).
10. S. D. Stookey and R. D. Maurer, "Catalyzed Crystallization of Glass in Theory and Practice," Tech. Papers International Glass Congress, Munich, 1959.
11. S. D. Stookey, J. S. Olcott, H. M. Garfinkel, and D. L. Rothermel, "Catalyzed Crystallization of Glass in Theory and Practice," Tech. Papers International Glass Congress, Washington, D.C. 1962.
12. S. D. Stookey, "Ceramics made by Nucleation of Glass - Comparison of Microstructure and Properties with Sintered Ceramics," Symposium on Nucleation and Crystallization in Glasses and Melts, Am. Ceram. Soc., 1962.
13. S. D. Stookey, "Modern Glass," Encyclopedia International Science and Technology, 1962.
14. S. D. Stookey, "Catalyzed Crystallization of Glass - Theory and Practice," in Progress in Ceramic Science, Vol. 2. Edited by J. E. Burke, Pergamon, New York, 1962.
15. W. H. Armistead and S. D. Stookey, "Silicate Glasses Sensitized by Silver Halide," Science, **144**, (3615) 150-54, 1964.
16. S. D. Stookey, "How Microcrystals Work in Photochromic Glass," Ceram. Inc., **82**, [4] 97-101 (1964).
17. S. D. Stookey, "Strengthening Glass and Glass-ceramics by Built-in Surface Compression," Proc. Berkely Intern. Mat. Conf., pp. 669-80, 1965.

18. S. D. Stookey, "Phase Transitions in Condensed Systems," *Molec. Design Mater. Devices*, 61-68 (1965).
19. S. D. Stookey and R. D. Mauer, "Glass," pp. 8-85 in *Handbook of Physics*, 2d ed. Edited by E. U. Condon and H. Odishaw, 1967.
20. R. J. Araujo and S. D. Stookey, "Photochromic Glasses, Properties and Applications," *Glass Ind.*, **48**, [12]687-90 (1967).
21. S. D. Stookey, "Selective Polarization of Light by Absorption Due to Small Elongated Silver Particles in Glass," *J. Appl. Optics*, **7**, [5] 777-79 (1968).
22. S. D. Stookey, "Chemical Principles Exemplified, Sunglasses that Respond to Brightness," *J. Chem. Education*, **47**, [3] 176.
23. S. D. Stookey, "Glassy State, New Frontiers; pp. 305-29 in Top. Modern Phys. Edited by W. E. Brittin, Colo Association, 1971.
24. S. D. Stookey, "Glass Chemistry as I Saw It," *Chemtech*, Aug. 1971, pp. 458-65.
25. S. D. Stookey, G. H. Beall and J. E. Pierson, "Full-color Photosensitive Glass," *J. Appl. Phys.*,**49**, [10], **59**, 5114-23 (1978).
26. S. D. Stookey, G. H. Beall, and J. E. Pierson, "Full-color Photosensitive Glass," *J. Phot. Sci.*, **26**, [5] 209-12 (1978).
27. S. D. Stookey, G. H. Beall and J. E. Pierson, "Full-color Photosensitive Glass," *Ceramic Ind.*, **110**, [6] 37-39 (1978).
28. S. D. Stookey, "The Pioneering Researcher and the Corporation," IRI Achievement Award Address, *Research Management*, **13**, [21] 15-18 (1980).
29. R. F. Bartholomew, P. A. Tick, and S. D. Stookey, "Water/Glass Reactions at Elevated Temperatures and Pressures," J. Non-Cryst. Solids, **38–39**, [2] 637–42 (1980).

U.S. Patents Issued to S. D. Stookey

1. 2,515,937, "Photosensitive Gold Glass and Method of Making It," July 18, 1950.
2. 2,515,938, "Photosensitive Copper Glass and Method of Making It," July 18, 1950.
3. 2,515,939, "Opacifiable Photosensitive Glass," July 18, 1950.
4. 2,503,140, "Glass Tube and Composition," April 4, 1950.
5. 2,564,978, "Multicellular Glass and Method of Making It," August 21, 1951.
6. 2,515,940, "Photosensitive Opal Glasses," July 18, 1950.
7. 2,515,941, "Photosensitive Opal Glasses," July 18, 1950.
8. 2,559,805, "Opal Glass Composition," July 10, 1951.
9. 2,515,275, "Photosensitive Glass," July 18, 1950.
10. 2,515,942, "Photosensitive Glass Containing Palladium," July 18, 1950.
11. 2,515,943, "Photosensitive Glass Article and Composition and Method for Making It," July 18, 1950.
12. 2,651,145, "Photosensitively Opacifiable Glass," September 8, 1953.
13. 2,682,134, "Glass Sheet Containing Translucent Linear Strips," June 29, 1954.
14. 2,628,160, "Sculpturing Glass (Basic Fotoform)," February 10, 1953.
15. 2,684,911, "Photosensitively Opacifiable Glass," July 27, 1954.

16. 2,651,146, "Method of Opacifying the Surfaces of Glass Articles," September 8, 1953.
17. 2,732,298, "Method of Producing a Photograph on Glass and Article Made Thereby," January 24, 1956.
18. 2,971,853, "Ceramic Body and Method of Making It," February 14, 1961.
19. 2,733,167, "Method of Adhering Gold to a Nonporous Ceramic Surface and Composition Thereof," January 31, 1956.
20. 2,921,860, "Opal Glass," January 19, 1960.
21. 2,779,136, "Method of Making a Glass Article of High Mechanical Strength and Article Made Thereby," January 29, 1957.
22. 2,911,749, "Method of Producing a Photograph in Glass and Article Made Thereby," November 10, 1959.
23. 2,920,971, "Method of Making Ceramics and Product Thereof (Basic Pyroceram)," January 12, 1960.
24. 3,157,522, "Low Expansion Glass-ceramic and Method of Making It," November 17, 1964.
25. 3,241,935, "Bone China and Method of Making It," March 22, 1966.
26. 2,960,801, "Method of Making a Semicrystalline Ceramic Body," November 22, 1960.
27. 2,933,857, "Method of Making a Semicrystalline Ceramic Body," April 26, 1960.
28. 3,253,975, "Glass Body Having a Semicrystalline Surface Layer and Method of Making It," May 31, 1966.
29. 2,998,675, "Glass Body Having a Semicrystalline Surface Layer and Method of Making It," September 5, 1961.
30. 3,195,030, "Glass and Method of Devitrifying Same and Making a Capacitor Therefrom," July 13, 1965.
31. 3,325,265, "Method of Making Synthetic Mica Bodies," June 13, 1967.
32. 3,135,046, "Method of Forming Metallic Films on Glass," June 2, 1964.
33. 3,249,467, "Method of Forming Metallic Films on Glass," May 3, 1966.
34. 3,208,860, "Phototropic Material and Article Made Therefrom (Basic Photochromic)," September 28, 1965.
35. 3,252,374, "Means for Controlling the Light Transmission of a Phototropic Glass Structure," May 24, 1966.
36. 3,231,399, "Semicrystalline Ceramic Bodies and Method," January 25, 1966.
37. 3,205,079, "Semicrystalline Ceramic Body and Method of Making It," September 7, 1965.
38. 3,268,315, "Method of Forming a Devitrified Glass Seal with Tungsten or Molybdenum," August 23, 1966.
39. 3,282,770, "Transparent Devitrified Strengthened Glass Article and Method of Making It," November 1, 1966.
40. 3,197,296, "Glass Composition and Method of Producing Transparent Phototropic Body," July 27, 1965.

41. 3,293,052, “Glass Article and Method of Making It,” December 20, 1966.
42. 3,449,103, “Photochromic Glass Making,” June 10, 1969.
43. 3,498,803, “Glass and Glass-ceramic Steam Treatment Method and Article,” March 3, 1970.
44. 3,498,802, “Steam Treatment Process to Produce Thermoplastic Materials and Hydraulic Cements,” March 3, 1970.
45. 3,656,923, “Method for Strengthening Photochromic Glass Articles,” April 18, 1972.
46. 3,540,793, “Photochromic Polarizing Glasses,” November 17, 1970.
47. 3,653,863, “Method of Forming Photochromic Polarizing Glasses,” April 4, 1972.
48. 3,653,864, “Dealkalization of Glass Surfaces,” April 4, 1972.
49. 3,782,982, “Products Prepared from Soluble Silicate Solutions,” January 1, 1974.
50. 3,827,893, “Silicate Bodies,” August 6, 1974.
51. 3,811,853, “Degradable Glass Suitable for Containers,” May 21, 1974.
52. 3,912,481, “Method for Making Alkali Metal Silicate,” October 14, 1975.
53. 3,948,629, “Hydration of Silicate Glasses in Aqueous Solutions,” April 6, 1976.
54. 3,940,277, “Glass-ceramics Containing Fibers Exhibiting Thermoplastic Properties,” February 24, 1976.
55. 4,017,318, “Photosensitive Colored Glasses,” April 12, 1977.
56. 4,057,408, “Method for Making Photosensitive Colored Glasses,” November 8, 1977.
57. 4,087,280, “Method for Enhancing the Image Contrast in Color Television Picture Tubes,” May 2, 1978.
58. 4,134,747, “Method of Forming Transparent and Opaque Portions in a Reducing Atmosphere Glass,” January 16, 1979.

CORNING GLASS’ PATENT FOR PYROCERAM UPHELD BY U.S. APPEALS BENCH

Lower Court’s Finding Reversed, And Corning Infringement Suit Over Anchor Hocking Remanded

By a WALL STREET JOURNAL Staff Reporter

PHILADELPHIA - The U.S. Appeals Court here reversed a Federal district judge who had found invalid a patent by Corning Glass Works, Corning, N.Y., for the manufacture of glass ceramic cookware.

Judge Gerald McLaughlin, ruling for himself and his colleague, Judge Harry E. Kalodner, in a patent-infringement case brought by Corning against Anchor Hocking Glass Corp., Lancaster, Ohio, said the district court in Wilmington, Del., had interpreted Corning’s patent too strictly in finding last March that it was invalid.

The appeals court upheld the patent and remanded the case to the district court to consider whether Anchor Hocking had infringed the Corning patent.

The patent covers glass ceramics produced by Corning under the trade name, Pyroceram, and includes the company's line of Corning Ware ceramic cookware, materials used in missile nose cones and building materials.

[In Lancaster, an official of Anchor Hocking said it hadn't yet received a copy of the appeals court decision and thus wasn't able to say what action it would take. The official said, however, that the ruling would have "no significant impact on Anchor Hocking's sales or earnings since the item in question is no longer carried in Anchor Hocking's catalogs."]

The district judge, Caleb M. Wright, had ruled that the patent was invalid because its claims were indefinite and required Anchor Hocking to incur too great an expense to show that it hadn't adopted Corning's process. Judge Wright didn't rule on whether Anchor Hocking had actually infringed the patent.

Judge McLaughlin found, however, that the 1960 Corning patent for crystallizing glass into a glass ceramic material was a "pioneer patent" for a basically new process and that the court was required to give it "a wide breadth of protection in construing the patent claims and specifications."

The point at issue involved the percentage of crystallinity in the ceramic glass and the method of measuring it. Judge McLaughlin ruled that Judge Wright had weighed the Corning patent too strictly and said: "It is quite apparent that the designation of 50% crystallinity was an effort to describe the conversion of a glass to a predominantly crystalline body and wasn't an arbitrary point at which the patentee (Corning) wished to assert a monopoly."

" It remains to be said," Judge McLaughlin added, "that the patent here found valid is a great basic invention which has brought substantial benefit to the public at large."

Chronological Table of Selected Stookey Inventions

Observation	Concept	Invention	Patent	Commercial Product
Need for continuous machine-drawn thermometer tubing. Failure of known opal glasses 1941	New zinc sulfide opal glass opacifies before the tube drawing starts 1943	First machine-drawn thermometer tubing 1945	1950	Thermometer tubing 1946
Dalton's opal patterns. Dalton's copper ruby glass (photosensitive) 1940	Develop photosensitive opal instead of thermosensitive opal in sodium fluoride glass. Copper ions in glass can undergo photochemical oxidation-reduction reactions 1941	Phot. opal experiments fail with sodium fluoride opal glass. Copper photo theory succeeds. Photo-copper glass perfected 1941	1950	
Gold glass develops ruby color on reheating like copper ruby glass, but is not photosensitive 1942	Gold in glass can undergo photoreactions, but requires optical sensitizer (cerium oxide) 1942	Photo-gold ruby glass 1942	1950	

Observation	Concept	Invention	Patent	Commercial Product
	Photo patterns of copper or gold crystals may cause (nucleate) growth of other crystals to produce photosensitive opals 1943	Photosensitive lithium Silicate. Photosensitive barium silicate 1944	1950 1950	
Heating, cooling and reheating after u.v. exposure causes gold particles to nucleate sodium fluoride crystal growth 1951		Photosensitive sodium fluoride opal glass, FOTALITE® 1951	1953	Lighting panels Windows Architectural spandrels 1958-1960
Need for color TV aperture masks, 1949. Discovery that phot. Li silicate opal glass is differentially soluble in hydrofluoric acid 1950	One of the phot. opal glasses may dissolve differentially where crystallized in some solvents Early 1950	FOTOFORM®, photochemically machinable glass	1953	Electronic products specialty microstructures 1953
A plate of FOTOFORM®, accidentally heated at high temperature, did not melt and was stronger than glass 1954	Photonucleated crystallization can change glass into a new polycrystalline material 1955	FOTOCERAM®, a hard strong photocrystallized material 1955		Electronic products, specialty, microstructure products. Decorative products 1955

(Continued next page)

Observation	Concept	Invention	Patent	Commercial Product
Li^+ ions replace Na^+ ions when glass above annealing temperature is treated in molten salt. Glass strength is increased 1956	Glasses containing alkali ions can be strengthened by ion exchange 1955	Glass strengthened by ion exchange 1955	1957	
FOTOCERAM® discovery 1955	All glasses can be controllably nucleated to form new polycrystalline substance 1957	Glass-ceramics, a new field of polycrystalline substances made from glass 1957	1960	CORNINGWARE®, 1958. Radomes, 1958. CENTURA® tableware, (J. MacDowell)
Clear glass conversion to clear low expansion glass-ceramic 1965	None. Accidental observation	Transparent low-expansion glass-ceramics 1965	1966	VISIONS® low expansion, transparent cookware (K. Chyung), 1980
Glasses containing copper- sensitized silver chloride crystals can be transparent, darken reversibly 1960	Subcolloidal silver chloride crystals in glass may cause glass to darken reversibly (Armistead) 1959	Photochromic glass darkens reversibly in sunlight 1960	1965	Spectacles. Sunglasses 1968

Observation	Concept	Invention	Patent	Commercial Product
Fluormica glass-ceramics were made and demonstrated to be machinable 1965	Soft crystals in glass-ceramic may result in machinable products 1965	Machinable glass-ceramics 1965	1967	MACOR® (G. Beall) 1978
Cups survived a 9-story drop to a steel floor 1960	A very low-expansion crystalline skin on a high expansion glass body may produce a very strong product 1959	High-strength articles. Glass with crystalline eucryptite skin 1960	1961 1966	
Steam treatment can incorporate water into some glass compositions 1970	Rubbery glasses may be made if water becomes part of the structure 1970	Self-degrading, hydro-glass containers 1973	1970 1974	
Pure silica glass and other glasses were made from silicate solutions 1972	Glasses may be manufactured from silicate solutions 1971	Low-temperature glass manufacturing process 1972	1974	
Pastel rainbow colors in FOTALITE® opal glass 1953	A complete spectrum of colors is produced by subcolloidal silver particles having different shapes 1975	Full-color photosensitive glass 1975	1977	

I.R.I. Achievement Award Address, "The Pioneering Researcher and the Corporation"

Ideas for innovation come from many sources. However, in these days of lament for decreasing innovation, it seems appropriate to consider how the corporation might increase the numbers and effectiveness of one idea source — its pioneering researchers who explore scientific frontiers for knowledge that can become breakthrough innovations.

In my paper I will first discuss some of the problems that keep most scientists from becoming true pioneers, or that decrease their efficiency, or divert them from the pioneering path, and then outline some thoughts on ways in which the corporation can help its pioneers and perhaps increase their numbers.

Figure 1 shows a picture of a mature pioneering researcher. The chief distinguishing feature is a heavy coat of armor on its bottom. Very few aspiring young pioneering researchers survive as such to maturity. Contrary to popular opinion, these birds have the riskiest job in a corporation, so risky that most scientists steer clear of it in favor of other careers within or outside the corporation.

For example, consider the fact that the explorer of a new frontier has no idea whether any treasure exists there; or if so, whether it is big enough or the right kind of treasure to interest his company. How long can he explore unsuccessfully before he is considered a failure? Two weeks, a month, a year, ten years? What if he discovers a silver mine, but his corporation wants only gold? Or, if he does discover a gold mine, will he be rewarded any better than his friend who works in a less risky job? The corporation should minimize the personal risk and maximize the rewards for success, in order to make scientists more willing to explore the unknown.

A Dangerous Job

We are all familiar with the well-known hunters who are making existence difficult for the corporation itself: inflation, recession, the risk-free society, pollution problems, etc. (Figure 2). When a corporation is forced to economize, its first instinct is to cut back on items such as long-range research that do not show up on today's profit and loss sheet. I submit that — wise management should protect the corporation's long term growth potential by maintaining its pioneering research, or even increasing it when national recessions are foreseen.

Most of these dangers — zero risk society, government regulation, inflation, recession — are self evident. The danger entitled university refers to such factors as the mediocre science teacher who turns creative minds away from science, or the research professor who convinces bright students that useful application of science is ignoble, diverting many creative minds from industrial research. The excitement and value and rewards of pioneering research should be effectively taught.

During the incubation period for a significant original idea, the main requirement is time for concentrated thinking and reading. Any interruption at this stage is a danger to the process of pioneering research. Thus meetings, fire calls to solve today's production problems, and assignments to improve on known inventions, are generally enemies to the birth of a new innovation (Figure 3).

The increasingly popular use of research teams certainly has its place, but original ideas come one at a time; but teamwork applies to the later stages of innovation and should not displace the individual inventor.

And as everyone knows, an embryo invention is a fragile flower, easily killed by the pessimism that seems to be a predominant characteristic of anyone over six years old. The pioneering researcher must be an optimist himself, and strong in character, to overcome the disbelief of his fellows in anything new. As research directors, you can save many inventions by at least pretending optimism.

A moderate sized growth corporation is probably the best home for a pioneering researcher. Very small corporations perhaps cannot afford to support one and capitalize on his inventions. Excessively large corporations have too much inertia, too many management layers, too impersonal organization, and require a multimillion dollar innovation to stir any interest.

Suppose now our young researcher, after working hard for one or five or even ten years, really succeeds in originating a blockbuster idea that culminates in an important innovation for his corporation. What is the probability that he will start over, searching for another new idea? This is another point of critical decision for the researcher. One or more attractive alternative paths for advancement may appear, competing with the lonely trail of the pioneer (Figure 4).

These other positions are all necessary to the corporation's welfare, relatively safe, and may have more potential for financial rewards and prestige than that of the successful pioneer. Depending on the inclinations of the individual, a position on an already existing project or in the analytical lab, or a position supervising others, may well be more attractive.

In order to be useful to the corporation, the pioneering researcher should be on the same wavelength as the corporate long range planners, and be aware at all times of the corporate hopes and its limitations in terms of product areas.

Support and Reward

As shown in Figure 5, the pioneering researchers should have the support of top management and research related departments. But before concrete support can be given to researchers, the corporation must first believe firmly in exploratory research, and, in effect, treat it like a maverick. To do this, top management must:

1. Give constant support, independent of business cycle.
2. Don't manage it — support it.
3. Recognize and reward exceptional risk-taking achievement.
4. Associate promising young researchers with successful pioneering researchers.

Following are two check lists that summarize ways in which the corporation can motivate and reward pioneering researchers:

How to Motivate Pioneering Researchers

1. Convince them the corporation needs and wants breakthrough technical innovation.
2. Convince them the corporation has, or is looking for, unexplored fields of technology.
3. Give them close association with active proven pioneering researchers.
4. Challenge them to stick their necks out, take risks, hit for home runs.
5. Reassure them that unsuccessful gambles don't mean catastrophe.
6. Reward successful risk-taking research appropriately.
7. Remove obstacles to long-term exploratory research.
8. Actively support researchers — equipment, technical services, lab assistant.
9. Encourage, don't discourage, original ideas.
10. Keep them up to date on corporation goals.

How to Reward Pioneer Researchers

1. Really try to use their inventions.
2. Recognize and reward them as risk-takers.
3. Grant deserved prestige within the corporation; recommend for deserved honors outside the corporation.
4. Grant independence (within broad corporation interests) in choice of research; and all reasonable support.
5. Grant special privileges (sabbaticals, etc.) if requested.
6. Financial: reward successful risk-takers appropriately by means of bonus, stock options, a dividend tied to profit from inventions.
7. Don't be afraid to reward the successful inventor. It might motivate others to try harder.

One Researcher's Experience

In 1940, Corning Glass Works was a medium-sized growth corporation, with $25 million in sales. Corning had already established a research lab more than 40 years earlier; and as early as 1859, Amory Houghton Jr., a son of the company's founder, had been doing glass composition research. Recently an event took place that emphasizes the important fact that the Houghton family continues to faithfully support research. The Amory Houghton Professorship of Chemistry was announced on October 15 by the President of Harvard, who stated, "The Amory Houghton Professorship honors a great believer in Harvard, in science, and the best in people."

Going back to 1940, I was placed under the guidance of two proven inventors, Harrison Hood and Bob Dalton, to learn about "opal glass" — actually, the science of nucleation and crystallization. At the same time, Dalton told me about a fascinating photochemical reaction he had discovered. Within a year, I was convinced that high temperature glass chemistry was a research field of unlimited opportunity, and had begun discovering new phenomena as an independent inventor. Each new discovery was of interest to Amory Houghton, who wanted to hear about it in person.

My roommate in those early days was Dr. William H. Armistead. He was a fine researcher and prolific inventor, but chose the executive path. Fortunately for me, he became research director. His know-how and sympathy with research led him to do everything possible to support our exploratory research. And his successor, Jack Hutchins, has continued in the same way.

Some of my early inventions (thermometer opal, photosensitive opal glass) brought modest financial return. Fotoform, a chemically machinable glass invented after ten years, was a delayed action innovation which attained breakthrough proportions only after another twenty years. Pyroceram, the first big innovation, was first commercialized 17 years after I started work at Corning. I think this demonstrates great patience on the part of Corning's management.

What has Corning done for me as a successful pioneering researcher? It has given me independence, prestige, and a position of friendship with Corporation officials; has rewarded me financially as well as it rewards key management people, including grants of stock options. In later years I was granted a sabbatical year at full pay, and subsequently the privilege of choosing my own working hours. And on retirement, my wife and I were guests of honor at an annual dinner of Corporation officers and directors where Amory Houghton Jr. praised my accomplishments.

The Industrial Research Institute Achievement Award to Stanley Donald Stookey

For his pioneering research in control of hitherto unwanted crystallization in amorphous glass; for his leadership and personal participation in creation of whole new families of widely useful glasses resulting from nucleation and crystallization of desired species in glasses: the photosensitive, photochromic and polychromatic glasses; for his vision in extending controlled crystallization to materials that are highly crystalline: the glass-ceramics. For his wise, thoughtful, and generous support and development of his associates within the laboratory and for his leadership in civic developments within the community.

Dr. Stookey joined Corning as a research chemist in 1940. He was named senior research associate in chemistry in 1950, manager of fundamental chemical research in 1955, and director of fundamental chemical research in 1963. Retiring in 1978, he is now consultant to the Corning laboratory.

During his research work at Corning, Dr. Stookey has been awarded 58 patents, was the author of 35 papers, and received numerous awards for his research and its results.

Dr. Stookey has twice received the Franklin Institute's John Price Wetherill Award (1953 and 1962) and in 1960 received the American Ceramic Society's Ross Coffin Purdy Award as the author who had made the most valuable contribution to ceramic technical literature that year. He also was given an Alumni Award of Merit (1955) and an honorary doctor of science degree (1963) from Coe College. In 1964, he was awarded the Toledo Glass and Ceramic Award.

In 1970, he was named Inventor of the Year by George Washington University, in 1971 received the American Chemical Society's Award for Creative Invention and the Eugene C. Sullivan Award from the Corning Section of the ACS, and in 1973 the Beverly Myers Achievement Award from the Educational Foundation in Ophthalmic Optics, and is the recipient of the glass industry's 1975 Phoenix Award.

Dr. Stookey was graduated magna cum laude from Coe College, Cedar Rapids, Iowa, in 1936. He received his master's degree in chemistry from Lafayette College in 1937, and was awarded his doctorate in physical chemistry from Massachusetts Institute of Technology in 1940.

He is a member of Sigma Xi and the British Society of Glass Technology; a Fellow of the American Ceramic Society and the American Institute of Chemists, and was selected to the National Academy of Engineering in 1977.

This paper was presented in October 1979 at the Fall Meeting of the Industrial Research Institute.

Reprinted from *RESEARCH MANAGEMENT, Vol.* XXIII/No. I (January 1980)

Awards Granted to S. D. Stookey

1. John Price Wetherill Award to the Franklin Institute (1953 and 1962).
2. Coe College Alumni Award of Merit (1955).
3. Ross Coffin Purdy Award of the American Ceramic Society (1960).
4. Coe College Hon. D. Sc. (1963).
5. Toledo Glass and Ceramic Award (1964).
6. Inventor of the Year Award, George Washington University (1970).
7. Award for Creative Invention of the American Chemical Society (1971).
8. E. C. Sullivan Award, Corning Section, American Chemical Society (1971).
9. Myers Achievement Award of Ed. Foundation in Ophthalmic Optics (1973).

10. Phoenix Award of Glass Industry (1975).
11. Achievement Award of the Industrial Research Institute (1979).
12. Samuel Giejsbeek Award of the Pacific Coast Sections, American Ceramic Society (1982).
13. Distinguished Inventor Award, Central New York Patent Law Association (1984).
14. Alfred University Hon. D. Sc. (1984).

Memberships

1. Sigma Xi
2. British Society of Glass Technology
3. American Chemical Society
4. Fellow, American Ceramic Society
5. Fellow, American Institute of Chemists
6. National Academy of Engineering (clected 1977)
7. Rotary Club of Corning (past member)

Acknowledgments

I am happy to express my heartfelt thanks to friend and colleague Doris Evans, Ph.D. and philosopher, for her suggestions and encouragement; to Joseph E. Pierson, my hardworking assistant and coinventor for the past ten years; to George Beall, Ph.D., for special friendship and fruitful collaboration; to Dr. John Hutchins, III and Dr. Harmon Garfinkel, Senior Vice President and Director of the Research and Development Division and Director of Research, respectively, for their friendly support of my post-retirement research and writing; and to secretaries Patricia Christian and Shirley Thomas for the many patient hours spent on the manuscript. My thanks also go to Ray Voss, retired Director of General Product Development, for converting many inventions into products; to Francis Hultzman for thousands of glass melts; and to David Morse, Ph.D., for his enthusiasm and success in carrying the invention process to a new generation. Most of all, I thank my beloved wife, Ruth, for her never-failing support and encouragement, and Dr. W.H. Armistead for his career-long friendship, inspiration, and support.

REFERENCES

[1](a) Antonio Neri, "L'Arte Vetreria," Florence, 1612; (b) C. Merret, "The Art of Glass," London. 1672; (c) J. Kunckel, "*Ars Vitraria* Experiments." Amsterdam and Danzig, 1679.
[2]W. Ganzenmüller, Glastech. Ber. 19. 325-30 (1941).
[3]Ibid, p.
[4]Ibid, p.
[5]W. Weyl, Colored Glasses, Sheffield, Soc. Glass Technology, 1951.
[6](a) A. W. Deller, Deller's Walker on Patents, Vol. 4; 65 0965); (b) Patents, Beirne Stedman, pp. 284-285 (1939).
[7]J. G. Vail, Soluble Silicates, Their Alkali Properties and Uses, Rheinhold, New York, 1952.

FLOW CHART OF SOME INVENTIONS FROM CONCEPT TO PRODUCT

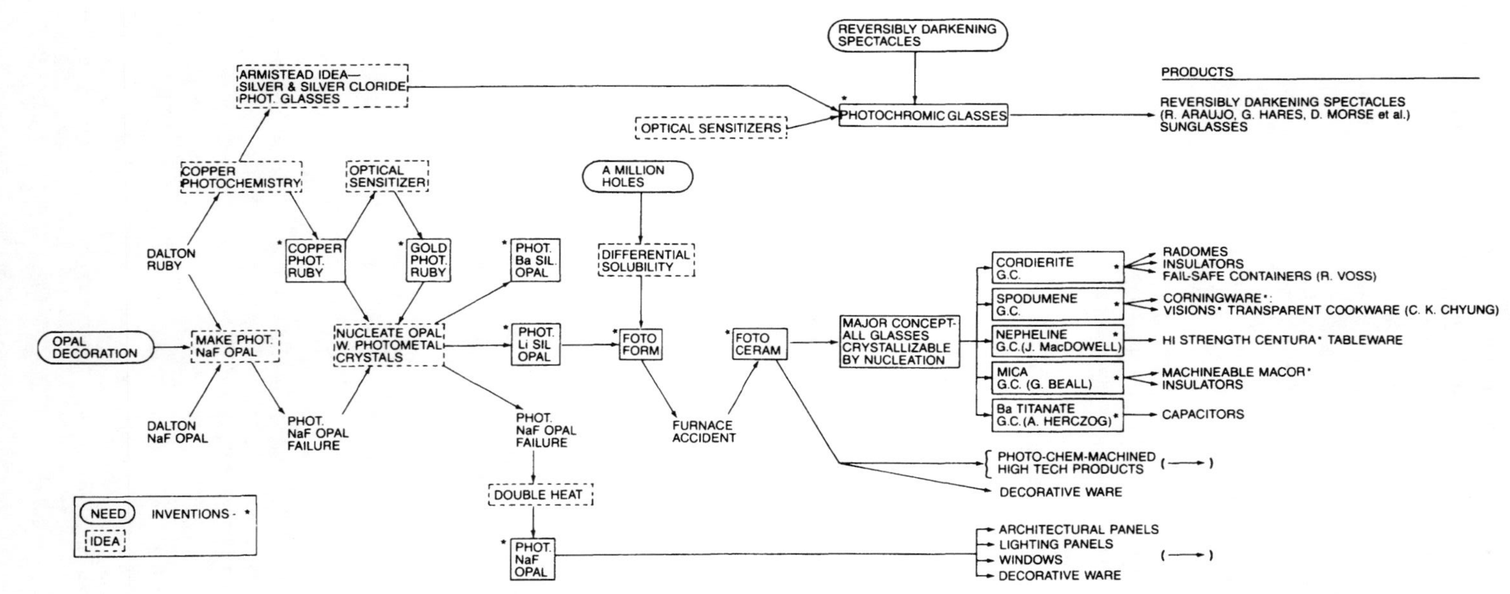

Printed and bound by CPI Group (UK) Ltd, Croydon, CR0 4YY

10/06/2025

14686700-0001